GENERATIONS

John Dooley

Burlington, Vermont

Onion River Press
89 Church Street
Burlington, VT 05401
info@onionriverpress.com
www.onionriverpress.com

ISBN: 978-1-966607-21-2

Library of Congress Control Number: 2025912378

DEDICATION

To my grandfather,
Orlando Aniello,
an immigrant,
a businessman,
and an employer.

TABLE OF CONTENTS

PROLOGUE

The mass adoption of green technologies, like solar power, electric cars, grid scale batteries, and heat pumps that we are offering worldwide use green power, not fossil fuels to maintain our most up-to-date lifestyles. We are looking to leave the lithium batteries behind for the future of electric power distribution.

Clean energy has a tipping point, and eighty-seven countries have generally agreed on the facts: If we as humans are to survive, it is necessary to decarbonize our future environment. We must control the carbon in our atmosphere by moving from fuel sources in the Amazon Rainforest to building heat pumps in the home. It will be the real challenge for this next generation.

If the United States continues to surge forward with wind and solar power, it will account for half of our generating capacity in ten years. Quicker forecasts are being predicted around the world. Therefore, to mass educate our young people, regardless of the temporary cost, a solution like the G.I. Bill prepared by our greatest generation after the Second World War has become a necessity.

We must not fail to realize that today's young people can continue to show the greatness of the previous generation, but we need to invest in them now.

Wind and solar technologies must replace fossil fuels. The more they are used, the cheaper they will become and less of an obstacle. We must be prepared for a change in our lifestyle—from our homes upwards. Waiting for federal edicts to relieve the situation is merely burying our head in the sand. There are entrepreneurs throughout our land, with minds like those in Edison's time, that must be stimulated to share our new life and help build a decarbonized society.

This book is fact-based fiction with characters who are challenging our present and the previously proposed future. A new generation of people facing a new environmental cataclysm.

I have written several books of various interests. They have included characters who have turned out to be popular. Those include: *Forever Blue*, which is about police reform, and *Dusty Lady*, which is on immigration and the fight to gain citizenship. *Generations* is about the financial need of the family farmer, dairyman, and the carbon in our atmosphere.

Several of these characters have found their way into this new book, and I hope that you will enjoy them again.

CHAPTER ONE

Cliffsides is a classy senior independent living home and was a gift from the city. It is now the home base for Abediah Tutt and Phoebe Sinclair, who, whenever possible, spend their mornings enjoying the large wicker chairs on the front porch reading the morning papers.

One morning, after they had a fine breakfast, they used the rest of the morning to survey the morning newspaper. Abbey would survey the paper and hand it to Phoebe, who preferred the society news and the general news while Abbey, a retired detective, preferred police and sports news.

Phoebe remarked, "Abbey, look at this. My longtime girlfriend's son, Tom, has just received a master's degree in science from Rensselaer University. There's a picture of them together. Mrs. Priscilla Carlton and Thomas Carlton, Jr. of Carlton Plateau Farm."

She came over to Abbey's chair and laid the paper in his lap.

Abbey commented, "He's a good-looking young man."

Phoebe replied, "I haven't seen him since he was a little boy when I was teaching high school. I see Priscilla regularly at the hairdressers, and she always brags about him and their daughter, Nancy. I'll call her on the phone and congratulate her when we go in."

Abbey said, "He must be a bright boy; science is a tough subject to master. I wonder what kind of sciences he has specialized in. His dad, Tom Sr., and his ancestors have lived on Plateau Farm for five generations. It's six thousand, four-hundred acres. When my boys were young, I drove them up to the farm to buy vegetables and apples in the fall. It's a well-kept farm. Tom Sr. died just last year. Farmers—they have such a hard life."

Phoebe said, "I always liked Priscilla; she was a grade ahead of me when we were in school, but we got along very well. I went to her birthday parties a couple of times; they were always at her house."

Phoebe returned to the paper in her chair, and Abbey went back to his sports news. Just before lunch, Phoebe and Abbey went up to their respective apartments.

Phoebe called Priscilla Carlton and told her she had seen their pictures in the morning paper. She congratulated her and her son on receiving his master's degree in engineering sciences.

Priscilla thanked her and invited her to their party the day after tomorrow. It would be taking place on the farm at seven-thirty in the evening.

Phoebe replied, "I'd love to come; could I bring a guest?"

Priscilla laughed, saying, "You mean Abbey? I've heard you were tied to a retired police detective. Of course, he can come."

Phoebe said, "Yes, Abediah Tutt—we call him Abbey—and he's all mine."

Priscilla laughed and said, "Good for you, girl. I'm looking forward to meeting him. You two are the talk of the town. Living at Cliffsides, with that beautiful view of the Atlantic on the east side and the entrance to Hidden Bay on the west side."

Phoebe said, "We'll plan to be there at seven-thirty. I'm looking forward to meeting Tommy again. I haven't seen him for years."

Priscilla said quickly, "We don't dare call him Tommy since his high school graduation. But he will be happy to see you then."

Abbey and Phoebe joined each other for lunch, and she updated him on her conversation with Priscilla. He noticed she was kind of self-reflective and asked, "What's going on with you?"

Phoebe answered, "I was surprised that she knew I was tight with a retired police detective. She also mentioned that we were the talk of the town. I wonder what she actually meant."

Abbey laughed and said, "Maybe she's heard of your episode with the snub nose thirty-eight you had at our police pistol range."

Phoebe responded, "Oh my God, I guess I'll never live that down."

Abbey laughed and replied, "Nonsense, be proud of your skill—you're the heroine of the Abenaki Reservation and the Forever Blue Laboratory."

Phoebe replied, "I wonder how she heard about that."

Abbey didn't remind her that it was in all the newspapers.

Phoebe replied, "Thank God I didn't pose for any pictures with my shoulder holster!"

Abbey leaned over and whispered to her, "I guess you haven't noticed we have very rapid service at the food store these days."

The next day, they were early to the party when Abbey drove into the parking lot of the farm's store.

Priscilla came out of the front door and shouted, "I'm glad you're early, Phoebe; I'm anxious to meet your famous retired chief of detectives."

She reached for Abbey's hand as she approached and shook it heavily.

Abbey smiled and said, "Thank you, ma'am. I've known Tom for many years. I've done business with you here."

"Thank you, call me Priscilla or Pris. May I call you Abbey?"

Abbey replied, "Of course."

Just then, a six-foot, two-inch, well-built young man walked up.

Priscilla took him by the arm and offered, "This is my son, Tom, Jr., and Tom, this is my longtime friend, Phoebe Sinclair, and her friend, Abediah Tutt—we call him Abbey."

Tom shook hands saying, "Glad to meet you again, Phoebe. I was only a boy when we last met."

Abbey interjected, "I followed your lacrosse career in the papers. Well, done. You've gotten a lot of good press."

Tom answered, "Thank you, but now I've got to get a job and make a living."

A voice everyone recognized came from inside the house and moved out on the porch. "He's got a job with me at my wind farm anytime he wants it."

Phoebe shouted, "Stamford Pettibone, what are you doing here?"

Stamford replied, "Pris is my sister; Tom's my nephew."

Phoebe looked at Priscilla and said, "You're a Castleton, not a Carlton."

Pris answered, "My father died when I was eight years old—Walter and Anne Pettibone, Stamford's parents, adopted me. Tom Carlton worked on their tobacco farm in the Connecticut Valley until his dad died. We married and came north to run the Carlton Plateau Farm. "

Abbey interjected, "Maybe Tom would like to market our Forever Blue Public Trust Police Body Cam Program and stay right here at home?"

Stamford agreed, "That sounds like a good idea. Let's hear what Tom's plans are. Though, it's still quite soon." He put his arm around Tom's shoulder.

Tom said, "Thank you, both. They're each fine offers. But I'm planning to add twenty to forty of our plateau acres to Agrivoltaic production."

Abbey looked at Stamford and asked, "What is Agrivoltaic?"

Stamford answered, "Solar panels that are fourteen feet above the ground with the farm crops growing underneath. It's the first time I've heard him mention anything about solar panels."

Tom continued, "I hope to add twenty to forty acres as soon as possible."

Stamford said to Abbey, "It's green power locally distributed on a generation grid. They are beginning to create them in this area and other parts of New England. It's part of that system, I believe."

Tom added, "It will add additional income to farmers because they'll get a percentage of the solar electric power sales to the grid."

Stamford then added, "The sales would be in megawatt—MW—power."

Abbey added, "I guess I'll have to study up on electric power and the business side of it."

Stamford then said, "I'll send you some literature for your homework. They're hoping to expand throughout New England."

Priscilla interrupted, "Okay guys, let's delay your business talks. Let's get on with our party; the other guests are beginning to arrive."

Phoebe took Abbey by the arm and led him into the yard beside the farm store. There were beautiful decorations on the poles, and the tables were set up with tablecloths, food, silverware, and a large cake with Tom's master's degree lettered on the top of it.

Tom was taken aback by Stamford. He was one of the honored guests. Stamford said quietly, "We have to talk seriously, Tom, about your future employment."

Tom responded, "Then think about Agrivoltaic, as well as sand batteries. Both of them can be packaged with the farm. I'd love to have you both supporting the whole project."

Mrs. Carlton started the party officially by introducing each guest as they arrived for the next couple of hours. Everyone had a fine time discussing Tom's past in school.

During the festivities, Stamford took Priscilla aside and told her what Tom had to say about his plans and then asked how she felt about it.

She laughed and said, "Like father like son. His father always strived for his dreams, but he never fell short of getting food on the table."

She added, "But I'd feel more secure if you and Abbey are close by during his initial efforts."

Stamford replied, "I'll ask Abbey if he'll join me in participating in Tom's projects."

Stamford followed through a short time later in a conversation with Abbey.

He said, "Let him follow his heart, Stamford. He's got enough good sense and an engineering background. We'll just make sure he has the financial and technical help where we feel he needs it. With the president's help, we'll have the Forever Blue Public Trust backing him up. Then he'll be used to us being around and talking things over with us."

Stamford agreed, "But you're going to be closer than I can be since I'm out to sea most of the time with my wind farm. It's fifty miles off to the North Atlantic coast. However, we can communicate regularly by the phone in the yacht that I keep harbored here at Cliffside."

Abbey agreed, "We'll do our best, and I'll start talking to the grid people in Cliffside and Bayside."

Stamford added, "Abbey, you have always been a forward thinker. That's why I backed Forever Blue Public Trust Corporation and your future project of body and cruiser cams. Together, let's fund the Carlton Point Corporation and its installation of the solar panels and sand batteries with Tom Carlton as Chief Operating Officer. I'll back it financially if you will be the treasurer, with Phoebe and Raphael Sullivan, our attorney on the Board."

Phoebe, who was standing not too far from Abbey's shoulder, said, "I'll be glad to cooperate in my position as the President of Forever Blue Public Trust Corporation. They are, after all, an organized group of multimillionaires financially willing to help nonprofit public service companies."

The party came to a natural end, and the guests started to leave. Stamford brought the point up again to get the reaction of the family about this decision. He restated his position about financial support before Mrs. Carlton, Tom's sister Nancy, and Tom's best girlfriend, Martha, left the party. They were one of the last guests to remain.

Tom said, "Hold on, guys. I can take chances, but you, as my friends, should not be bound by them."

His mother interjected, saying, "Tom! Stamford, Abbey, and Phoebe were not born yesterday. They are serious business people. They know, very well, what they're getting into. They know their minds, and I think it's a fine idea and a good backup to have."

Martha jumped in, saying, "I'm in favor of our business being here at home."

Phoebe laughed, took Pris by the hand, and said, "Do I detect an assumption of marriage?"

They all laughed, and Tom said, with a grin, "Trapped. I walked right into a trap."

He then looked straight at Stamford and Abbey and said, "Okay guys—you asked for it. As the Chief Operating Officer of this soon-to-be-created corporation, I want you to make a deposit to purchase the closed Cliffside Coal Plant immediately. As soon as tomorrow. In school last year, I wrote my master's thesis on how buying old coal plants helped solar power farmers in the Midwest become established. They purchased and modernized old coal plants for new uses on a green power grid. That plant will be the secret route to the local distribution green power grid. We're going to need it."

Stamford said to Abbey, "Who can we see in this community who knows the price of the old coal plant, as well as a realtor to close a deal?"

Abbey said, "It's Hawthorne Realty, owned by Henry Hawthorne, an old friend of mine. He should be willing to help the Forever Blue Public Trust Corporation buy the plant. He used to be a cop before he developed heart troubles; I know him well."

Stamford then said, "I'll pick you up here in the morning and we'll go see him. We'll drive by the old coal plant and size it up. Bring along a blank Forever Blue bank check, just in case we make a quick deal."

Abbey replied, "I'll be here at nine a.m."

Abbey was at their door at nine a.m.

Stamford got in the car and said, "Good morning, I had a nice talk with Tom last night. He's obviously given this move into Agrivoltaic and sand battery technology a lot of thought in order to support the local grid. He sees a good future with a green private power corporation. His general idea is to prepare the Cliffside and Bayside communities in the event that the National Grid cannot supply the local electrical load required, like during an electrical emergency."

Stamford continued, "Getting a local contributing distributing company to be accepted by the existing plant grid—even though the coal plant is probably still registered with the grid—will be a political battle, no doubt. However, according to what Tom has read in school, the big power grid guys are buying up old coal plants to be able to do just that. Tom is hoping that we can do the same."

Abbey answered, "Sounds like a good idea. If we can demonstrate that he has the financial backing and technical help that he needs to be able to play with the big boys. Stamford, did you have any difficulty getting the big boys to accept your wind farm?"

Stamford answered, "No, but I was already playing in the big boys' club with fossil fuels. I quit fossil fuels because I don't believe they have any future. They gave me a hard time, but the climate control people helped me create my wind farm fifty miles off the Atlantic Coast. The local green power people should be willing to help us as we get going."

Stamford added, "If we get this plant and twenty to forty acres of solar panels from the farm, we should be able to muster up some local help."

Abbey drove into the old northwest corner of Cliffside and came to the block where the old coal mine was. It had a Hawthorn Realty "for sale" sign out front.

Abbey parked the car on the street, and they walked along the fence on the property. It appeared to be well-sealed and locked up, including the outside parking area, which took a whole city block. The neighborhood had been well kept with small private homes, and the northwest area represented a quarter of Cliffside's 45,000-person population.

Stamford commented, "This is a pretty big plant for Tom's small business."

Abbey answered, "Dressed up, it could be a quiet complex of corporate offices, with solar panels on the roof to provide power and potentially power the whole neighborhood!"

Stamford said, "Okay, let's go talk to the realtor. Where is his office?"

Abbey replied, "Only twenty minutes away in the center of town, close to the police station."

They got back in the car, and Abbey took off. In a few minutes, they pulled into the realty parking lot. A tall, blonde, middle-aged man came out of the building. Abbey got out and said, "Good morning, Henry."

The man offered his hand and asked, "What are you doing here—are you on police business?"

Abbey shook his head and responded, "Real estate business. Stamford here is interested in the old coal plant."

Abbey then introduced Stamford to Henry Hawthorne. They shook hands, and Abbey continued, "Have you heard I'm connected to Forever Blue Public Trust corporation? Stamford and I are on the same board."

Henry responded, "I have heard. The cops aren't talking about anything else but their new police bodycams. How can I help?"

Abbey answered, "Forever Blue is interested in the old coal plant you have for sale. We want to turn it into a corporate business location."

Henry answered, "I'm glad someone wants to do something to fix it up. I know the neighborhood people have been hoping for that for some time."

Stamford asked, "What are they asking for the property?"

Henry answered, "The town wants a hundred and fifty thousand."

Stamford responded, "That's a lot of money for a building in that condition."

Abbey added, "Henry, you and I know that place has been falling down for years."

Henry looked at him carefully and said, "Yes, I know. I've been trying to talk them into something different, too."

Abbey said, "Call them on the phone and tell them we'll offer a hundred grand right now."

Henry said, "Have a seat in my office, and I'll call them directly."

He led them into his office and offered both a chair then went into another office to use the telephone.

Stamford whispered, "A hundred grand! What a steal."

Abbey nodded quietly. Henry came back with a strained look and said, "They agreed to a hundred thousand, but you have to pay ten thousand in back taxes."

Abbey looked at Stamford and said, "Okay, we'll go for it."

Abbey reached into his side pocket and withdrew his wallet, laid it on the desk, and wrote out a check for $110,000. He then turned and handed it to Henry.

Henry said, "Great! I'll have my secretary print up the papers right this minute."

Henry looked at the check, smiled, and said, "You guys relax, I'll be back with the papers as soon as she's finished typing them."

Stamford, smiled, and said to Abbey, "He'll take his wife out to dinner tonight; his commission ought to be about six percent."

In ten minutes, Henry was back, and he sat down at his desk and offered them the papers to sign. Abbey signed all the copies, folded their copy, put it in his pocket, and Henry gave him the keys. They all shook hands and left.

Abbey and Stamford climbed into Abbey's car and drove back toward Carlton farm. While in the car, Abbey said, "Stamford, take the papers and give the keys to Tom as a graduation present from his uncle."

Stamford said, "Thank you, Abbey."

They went into the house—everyone was in the kitchen. Phoebe had come over in her own car and had told Pris, "I just can't miss this."

Stamford walked over to Tom and handed him the papers and the keys while saying, "To the next generation. A small graduation present from your uncle Stamford."

Tom looked at the papers, choked up, and said, "The coal building is ours, and only for a hundred and ten thousand."

He got up and put his arms around Stamford. His mother wrapped her arms around both of them. Phoebe pushed back the tears and kissed Abbey, saying, "You're all invited as my guests for dinner at the Waterfront Tavern this evening. Nancy, will you call and make the reservation for the silo dining room? The table will include Tom Carlton, Mrs. Priscilla Carlton, Nancy, Carlton's uncle, Stamford Pettibone, Phoebe Sinclair, and Abediah Tutt."

CHAPTER TWO

They were all gathered on the steps of the Waterfront Tavern on the bayside at seven-thirty p.m. They enjoyed the view of the harbor and the ships at anchor when Nancy said, "I've always wanted to eat here."

Her mother said, "Your father brought me here years ago, but we sat at a table in the bar room; this is going to be impressive."

They went up the stairs and into the Waterfront Tavern and looked around as they climbed to the second floor. A waiter approached Abbey and told him they were reserved for the silo dining room. The waiter showed them the way and helped them get seated.

Abbey invited all of them to check their menus. The waiter would be back shortly to take their orders. In the meantime, he described some of the meals, translating the colorful titles, and Phoebe added some information about how the meals were prepared. The waiter returned and took their orders carefully and then left, thanking them.

Tom said, "Everything has happened so fast, I don't know where to begin."

Stamford answered, "I know how you feel. I felt that way when I first got the papers for the wind farm and the papers for the yacht in Peru at the same time. When I explained being overwhelmed to the chairman of my board, he said, 'If you're really overwhelmed, we'll throw you in the ocean and bring you back to life.' Everyone laughed back then, but I didn't take him up on it."

Tom said, "Okay, what should I do first with the coal building?"

His mother said, "Let's call some architects, take them through it as is, and tell them what you envision. They'll take pictures then give you some drawings as well as the immense estimate for cleaning up that dirty old plant and its offices."

Abbey added, "That will give you plenty to think about based on their expertise. If you approve, the Forever Blue Public Trust Corporation will encourage the architects to hire contractors. One company will be cleaning the building that was the coal operation, and the other cleaning company will deal with the offices. Send Forever Blue the bill, and I'll be happy to pay for it."

Stamford asked Abbey, "Why not use those architects who drew up the Forever Blue building next to the Algonquin University? They did a pretty good job, and they weren't all that expensive compared to other architects."

Phoebe spoke up and said, "Good idea! Tom, call the Wickiup architects in Boston and introduce yourself so you can explain the project you have for them. They'll call back to Forever Blue and verify the job. Just make an appointment to walk through the building with them."

Tom questioned, "Wickiup? That's a strange name for an architectural firm."

Abbey answered, "Yes, they're all graduates of the Algonquin Indian University. We gave them a lot of business then, and we'll give them a lot more now. They're very good."

Priscilla asked, "Are they all Native Americans, Abbey?"

Phoebe answered, "They're all graduates from Algonquin University, the free university for Native Americans right here in Cliffside."

Tom said, "I like that. I'll also get in touch with the Agrivoltaic people for an estimation of the cost of a twenty-acre panel installation. We'll take on a twenty-year lease starting in the springtime. I'll mention your name, Abbey, to the architects and to the Agrivoltaic people, okay?"

Abbey responded, "Yes, be sure to do that. They won't worry about their money. Consider it a favor to the city of Cliffside. The northwestern neighborhoods have been going downhill since the coal plant closed. I'd like to see the newly modified office complex provide jobs and a newly installed panel heat in their homes. In addition to the twenty acres at the farm, I hope the new energy source for them will be generated at the same plant. How do you feel about that, Tom?"

Priscilla added to that point, "Tom, your father always felt our farm was close to the people in the northwestern neighborhoods. He used to give them surplus farm products. Do you remember?"

Tom replied, "Mom, of course I remember. I delivered them many times myself. I'll have to make a large sign to advertise new jobs available for the local people. Stamford, maybe call your new corporation, Carlton Farm Enterprises, so that the people will know it's the Carlton family farm behind it."

Tom looked at his mother then at Nancy and said, "How do you feel about that?"

Pris looked at Nancy, smiled, and said, "We agree. Tell the lawyer, Raphael Sullivan, we are Carlton Farm Enterprise Incorporated."

Stamford mused, "I'm happy to be in relation to your new corporation for green power."

Tom said, "At school, we studied coal-fired electric plants and their shortcomings. One of the worst dangers is from the ash created from the process of converting the coal. It's going to be important to make sure our plant is clean before we attempt any new construction. We are converting coal to solar panels with a heat source involved in our process. We must clean the plant of any remnants of pollutants from the coal first.

"Therefore, it's necessary to have two different types of cleaning companies. One industrial company that is used to doing the heavy cleaning of the dirt within the plant, cleaning the bricks, and, if possible, the ash on the outside. The industrial company will be trained in dealing with potential pollutants. The second company is of a typical cleaner or sort and is used to cleaning operating office buildings. Hopefully we won't need a firm that can deal with any radioactive problems. Let's pray that we don't find any."

Phoebe said, "I can see you've studied the problem, and I agree it's wise to proceed carefully. As the president of Forever Blue, I'll clear the move for such action. As board members, Abbey and Stamford, do you agree?"

Abbey said, "Aye," and Stamford nodded his agreement as well without any hesitation.

Abbey added, "I've been reading national magazines concerning police actions in relation to the power grid. There is a possibility that bothers me. Here's a quote from one of the reports:

"Incidents related to vandalism, suspicious activity, and cyber events are on track to be the highest reported crimes since the reports started being kept. Increased rates of activity have been reported throughout various police departments around the country. Most prominently in Florida, Oregon, Washington, North, and South Carolina.

"According to some reports, the grid system is inherently vulnerable as it's spread wide across the country. This makes the lines and substations easy targets. I read that in February of this year, three men who were described as white supremacists and neo-Nazis pleaded guilty to federal crimes related to attacking the grid substations. Since we are going into the business, I suggest we also make a real effort to have our plant secure from such criminal efforts."

Tom thought for a moment and then said out loud, "I think you're probably right, Abbey. I haven't thought about that, but I'm glad that your police mind is working in that direction; it could be a serious problem. I have heard that in Cliffside we have a white supremacist group, am I correct?"

Abbey responded, "Yes, we do have a small militant group of white supremacists in Cliffside and Bayside, but they haven't been active and our detectives have stayed close, so I don't think we have to worry about them at the moment."

Tom answered, "Ash from the coal will be the main villain, then. A complete radioactive survey will be done afterward. When I talk to the Wickiup architects, I'll go over it. Then they will be prepared for the work. Stamford, do you and Abbey want to be there at the same time?"

Abbey said, "I'll be with you when you're in the plant and will bring some police back up. That'll keep us ahead of any potential problems with the law. Also, it'll give us some protection against nosey neighbors who are curious about what we're doing with the plant."

Stamford added, "I'll be there, too."

Tom reminded them, "It's important that the heating equipment and the electrical grid system be preserved for our future. It could be used by our enterprise. If we are not careful, it'll throw a monkey wrench into the wheel."

Stamford added, "As a wind farm owner, proper connections to the grid are of the highest importance."

Then the waiter arrived with a cart filled with their dinners. He passed them around with great care, giving them to the right person each time. He asked Abbey, "Shall I pour the decaf coffee now, sir?"

Abbey responded, "Yes, please."

The waiter proceeded to fill all the water cups from a silver urn. When he finished, he refilled the urn and placed it on the table next to the cream and sugar. Abbey thanked him and gave him their dessert order for later. The waiter nodded and left.

Priscilla then said, "Please, let's postpone the business talk and just eat."

They all agreed and went to work on their meals. In a few moments, Abbey looked over to Nancy and asked, "Nancy, does it live up to your expectations?"

She answered between mouthfuls, "It sure does. Thank you, Phoebe and Abbey."

Phoebe responded, "You are very welcome."

She slyly rubbed Abbey's leg. After the meal, they got their dessert and great compliments were given from filled stomachs. Abbey put his Visa card on the waiter's check tray and complimented him and the house on the meal. He returned their thanks and left with the tray.

Tom asked Abbey if he would make their corporate appointment with Raphael Sullivan. Abbey answered, "I'll be glad to, and we'll meet you at the bank building, then—if that's okay with you. I hope you'll have appointments with the architects by that time."

Priscilla suggested, "We've had a fine time and a great dinner, but I think it's time we took ourselves to bed."

They all nodded in agreement and said goodnight to each other as they made their way out the door. Abbey signed the receipt from dinner, and they all walked out into the fresh air.

The next morning, Abbey called Raphael and set the appointment at ten a.m. The following day, he called Tom and confirmed the meeting with him.

All three Carltons and Stamford were in the parking lot of the bank, waiting for Phoebe and Abbey the following morning. Abbey let them into the bank and up the elevator to Raphael's office.

Abbey introduced the Carltons to Raphael. Stamford said, "Good to see you again."

Raphael encouraged all of them to take a seat around a large table. Then he remarked that he was ready with the paperwork. He called his secretary, who placed all of her papers in front of the Carltons. She said, "If you have any questions, please ask."

She showed them where to sign next to the red arrows that she had provided on each form. Rafael started with Mrs. Carlton, Tom, and Nancy. Followed by Stamford, Phoebe, and Abbey to represent the Forever Blue Public Trust Corporation. Rafael then announced, "The Carlton Farm Enterprise Corporation is now a reality."

Raphael said, "I'll clear all of the papers with the State of Delaware and send them copies as well as copies to you."

Tom interjected, "I called Wickiup architects and introduced myself as a friend of Abbey and Forever Blue. They were very receptive. I explained I wanted to create a green electric power center and was looking forward to their coming out and spending some time examining how to clean up the property and to take pictures. They suggested a weekend as being more appropriate so that they could spend as much time as they need if they wish. I agreed, and they gave me a weekend two weeks from now. Is that okay with you all?"

Abbey looked over to Stamford and said, "It's okay for me, but you have to come from the wind farm, is that convenient for you?"

Stamford answered, "A weekend is more convenient for me than during the weekday—if I can come on Friday night and stay here at the farm. What do you think, Priscilla?"

She nodded. "Anytime, Stamford."

Stamford then said, "Tom, tell the architects, 'Okay.' You, Abbey, and I will meet them at the farm."

Tom added, "Fine, I'll set it up that way. I ordered the sign for the front of the building. I'll have them set that up inside the gate on the same Saturday."

Abbey added, "I'll arrange for police coverage that weekend to make sure we're not disturbed. I'll also schedule a drive by from the local cruiser so nothing will be disturbed."

Tom said then, "That's great, thank you both. It should go well. I'll tell the architects to take many pictures."

Stamford added, "I think they'll take pictures. It's a big part of their investigation, Tom."

Phoebe added, "It looks like you guys will have a productive weekend."

The next morning, on the porch, Phoebe asked Abbey, "How much do you know about the coal burning plant you just purchased?

Abbey replied, "Not too much. It was built in the 1920s when the country changed from gas light to coal-generated electricity. Cliffside had about 20,000 people at the time and has since grown to 45,000 because of the new electricity in the homes and the businesses.

"I was just a rookie cop then and was occasionally assigned to the northwest part of the town. My partner and I drove up the steep road that's north of town, where the road levels out, just before the Carlton farm.

"After another half mile, we came to a road called Factory Place. It was where they had built the coal plant, just a hundred yards south on a level spot. You can't see it now, because of all the houses, but we drove past that and over the roaring brook bridge to the town line road, where we turned south again into the town.

"There weren't too many houses around at that time, just woods between North Road and the rest of the town. The town road followed the roaring brook south and then east to the level acreage where they built the coal plant on the north side of the brook. From the North Road, we could only see the roof of the coal plant. As we came close, we noticed that the brook ran close to the rear wall of the plant and was used, in some way, by the coal plant itself before it continued.

"As we continued south, we cut back and forth through new streets on our way back to the police station located in the center of town. I usually had that route on duty one week a month."

Phoebe offered, "Our family lived a block north of the green in the center of town. When I was invited to one of Priscilla's parties, we would go up to Carlton Farm, but she had to ride the school bus. I never got out to the high part of town unless my dad drove us in the car."

Abbey continued, "The coal plant hired a lot of people who moved near the plant during World War II. Many families are still there. I have never been inside the coal plant, so it'll be an interesting trip for me."

Phoebe answered, "I think I'll wait until it's fixed up. I don't have any knowledge about electricity or coal plants. I just turn the switches on or off, and I know I can't live without it."

Abbey agreed, "I'm with you, but at the Police Academy they gave us more understanding on the topic in our training. At the station we had a backup generator, which runs at a certain desired speed. I think it's sixty cycles per minute. That generator has a governor, which is a control item, that phases fuel through a generator to maintain the power necessary to handle our load, depending only on the amount of power needed to run the electric devices we have connected. Also, there are protections built in to keep the generator from putting out too much voltage. They told us at the academy that if the generator continues to produce a thousand watts of power for an hour, it has produced one thousand-watt hours of energy. Not all of it is clear to me. But in other words, when power is produced for a period, it is multiplied by the amount of time the energy is produced. Also, for instance, one thousand-watt hours is the same as one-kilowatt house (kWh). That's what it says on our electric bill, so the number of kWh is what you pay for. They also confuse the reader by referring to it as a kilowatt load. I think they just use that jargon to confuse the rest of us.

"However, to play devil's advocate, by keeping the generator running, we found it avoided blackouts in individual areas. The more the city grew, the greater the need for more generation of power. Therefore, two of the small electric companies joined together to produce the power that Cliffside needed. They burned the coal to fuel the generators to make electricity.

"As the public's demand for electricity grew, the suppliers joined forces to save money as well as give them a better potential for the future need of electricity. They formed an organization called 'The Grid.' It had its risks, so they moved the plant further away from the center of town. Overall, it seemed to be a popular move."

Phoebe answered, "Okay, okay. I already believe most people need to know all that jargon when they buy a new appliance. "

Abbey laughed, "Okay, enough already. But you and I are seventy-five years old. The young people are getting all this information at a young age in school."

Tom, Stamford, and Abbey were ready on the agreed weekend. Peter Whitehorse, president of the architectural firm Wickiup, and his team arrived at the Carlton farm with all their equipment.

Tom introduced himself and shook hands with Peter Whitehorse and then introduced Stamford and Abbey.

The group left for the old coal plant on Factory Place.

The sign company's truck was already there, ready to install the new

Carlton Farm's Enterprises sign. It was four feet by six feet and had two steel poles to hold it up behind the fence that stood four feet high. Tom read the sign and said, "Good, unload. We'll open the gate, and you can install it right away. Do you agree, Stamford?"

Stamford said, "As the finance director, I agree."

Tom put his arm around Peter's shoulder and said, "You are all welcome."

A police cruiser arrived shortly afterwards. Abbey went out to greet them and arranged to have someone cover the gate for three to four hours while the architects' crew and they went through the building.

The sign men quickly finished and left. Several curious people walked up to read the sign. The sign seemed to satisfy their curiosity because no one asked any questions. One of them wrote down the phone number in case he needed any further information.

Tom led the group up the stairs to the front door, found the right key, and led them into the front lobby of the building. There was a reception desk and a chair in the center with a staircase behind it. Two officers stood on the left and two on the right of the staircase.

Abbey and Stamford walked around to check out the offices. They each had desks, chairs, filing cabinets with folders inside, blackboards on the walls, and two side building windows.

Peter asked Tom, "How do we do this?"

Tom responded, "Have the architects start taking pictures and measuring in here first then out in the main plant on this floor before we move upstairs or down into the basement. Have them also take photographs as we go.

"We're going to hire two cleaning companies. One industrial cleaning company for the plant that's accustomed to cleaning industrial sites. They'll work inside and outside of the building. Another company accustomed to cleaning offices will come in after.

"Get individual pictures of all equipment, and identify their uses if possible. We're going to be putting solar panels on the flat roof, so get some good shots there and outside as well."

Peter said, "Okay, Tom."

Peter walked over to his men and conveyed all Tom had told him to them. He then assigned different people to different jobs of taking pictures and measuring.

Tom motioned to Stamford and Abbey to follow him. "We'll let them do their thing while we wander around to satisfy our own curiosity."

Stamford and Abbey nodded. Abbey said, "Sounds good to me."

Tom added, "Let's go out into the plant first; I'd like to see what machinery or electric equipment has been left. We're looking for things such as turbines, heating equipment, and generators. It'll give us a feel for how things have been used and help us find connections with the brook. The same goes for the coal storage bins and conveyors."

Stamford agreed, saying, "I'd like to see how close the machinery is to what I use in the eight towers on my wind farm."

Abbey added, "You both can explain it all to me. My only experience is with the police emergency routine and the emergency gas-powered generator."

The rooms in the factory building were large with ten-foot ceilings. There was a wide staircase that led to the second floor. There were workbenches with connecting revolving belts overhead and wooden control arms coming down.

Stamford said, "A lot of the plants during the 1920s used overhead belts. We don't do things in the electronic age like that anymore."

Tom agreed, "Especially that big potbellied iron stove used for heat in the winter—or that rotating stone for sharpening blades."

On the east end, there were grinding machines to crush coal to dust before putting it into the water that turned steam to heat. The steam would pass through a furnace to power a turbine and its attached generator.

Tom said, "The principle is still the same: Water must be channeled from the brook rushing from the back of the plant. Let's go downstairs and see how water comes in and to the back of these machines—it should all be done in the basement. It was a piping system from the brook through the wall with a simple slide control. I'll bet someone got wet working that lever. We'll have to improve on that."

Stamford, looking over Tom's shoulder said, "We don't get that wet out on the Atlantic."

Abbey said, "I'll call the power company and have them change the brook outside to the plant's address. I imagine they'll have it all set in forty-eight hours."

Tom said, "Good, have them create an enterprise account. Now, let's see what's up top of the stairs."

They trudged up the wooden staircase and found a large open room to the west and what appeared to be a smaller room to the east with a big double door. Tom walked over and pushed one of the double doors open. He exclaimed, "Wow, look at this."

They all rushed over, pulled on the other door, and looked down into a giant pit. There was a short platform directly in front of them with a railing.

Abbey explained, "This must be the room where they dumped all the coal. It seems to be a giant coal bin."

He took hold of a rope and pulled hard.

CHAPTER THREE

A large door on the opposite side wall opened widely. Stamford said, "That's how the coal got in; they must have backed up trucks to get all that coal in here."

Tom added, "They must have used gas-powered conveyor belts to bring it up from a truck. That sure is a big coal bin. I wonder, what are the actual dimensions?"

Peter, who had come up to see what the commotion was about, said, "We'll know soon; we're measuring everything, including this."

Tom put his arm around Abbey and said, "You're looking at the new sand battery pit—the first in New England."

He then turned to Peter. "I want this pit clean from any coal ash and insulated to hold heated sand up to a thousand degrees."

Peter asked, "Heated sand?"

Tom answered, "Yes, it will power turbines and generators that will alternate DC and AC current and satisfy the loads that provide heat and power to the whole neighborhood. Look up the Finnish sand battery, that will explain it to you."

Stamford, looking at Peter, said, "The first in New England and maybe even the USA. Did you find the grid connections?"

Peter said, "Yes, on the first floor on the west side. There's a big set of controls, thirteen in all."

Tom added, "Peter, you must guarantee absolute and complete confidentiality."

Peter answered, "You got it. I'll supervise all the industrial cleaning inside and out and all the office cleaning so we can guarantee confidentiality. I'll get back to you with the cost and how long it'll take."

Tom said, "Peter, let's get those red bricks on the outside of this building. Our main challenge is coal ash; it's radioactive. It can kill the workers, tenants, and the program."

Peter said, "Tom, Stamford, and Abbey, we won't let you or the program down. I'm curious, though, about the rolled-up papers that you've kept under your arm?"

Tom smiled and took the roll from under his arm. He spread it out on one of the workbenches and waved the crew to come see.

Tom explained, "These drawings I made envision how the plant will be developed into a solar panel, hydroelectric, and hydrogen green power facility. The sand battery in the pit will provide storage for the solar green power."

Peter asked, "How does the sand fit in?"

Tom whispered, "My secret weapon is the sand battery."

Peter whispered back, "Batteries are not made with sand."

Still whispering, Tom answered, "We're going to develop a new method of heat generation for electric storage."

Peter answered, "I see. You're going to further develop the method the Finnish have just created."

Tom looked surprised and said, "I'm glad you keep up with scientific news."

Peter smiled and said, "It's not just me—a thousand American engineers read scientific magazines."

Tom answered, smiling, "But they don't have their own coal plant and solar Agrivoltaic farm array. Keep it to yourself."

Peter answered, "Wickiup never discloses our customer's business."

He then said to his partner, Greg, "Take pictures of all of these drawings."

Stamford and Abbey studied the drawings carefully then looked at each other but said nothing.

Greg spent some time taking several close-up pictures and then said, "We'll need some supplemental inside shots once the cleaning is done."

Abbey turned to Tom and said, "You're putting together a very top secret up-to-date electric organization. I'm a policeman asking questions for the benefit of the public, which I'm sworn to protect, so, as a policeman, I'd like to know your motivation. Is it financial? That's a popular drive for young people these days. Is it to build a business to ensure employment for the poor? Or a grand desire to help our working class? Or are you trying to save our planet from climate abuse?"

Tom thought a moment then motioned to the men around him and said, "Abbey's questions are fair ones. He has taken a Police Oath to protect the public as a policeman. If you're going to devote time, talent, energy, and money to this project, then you should know the answers. I'm a trained electrical engineer. I've learned that the American electrical grid system is a hodge-podge of private and public and out-of-date skeletons waiting to be destroyed by nature and ambitious people. The industry is under continued assault by terrorists or foreign computer geniuses competing for world power. I know that I cannot change the National Grid System, but I do believe we can help to save our city of Cliffside in the event that the nation's grid is attacked, and we experience a blackout.

"If we're successful, we will have an independent backup service prepared and ready, waiting to keep our community and its people with their power intact.

"Hopefully other communities and entrepreneurs will do the same. Our energy distribution over previous generations have used fossil fuels for power. Cliffside, if it is properly organized, will be prepared with green power and backed up with our sand battery and hydro-electric power to cover Cliffside's emergency needs—like police, fire, EMT, and immediate medical needs. Our community will be able to continue functioning until the New England Grid has been restored."

Abbey said, "You've got a powerful motivation that fulfills our Police Oath demands. We'll help you put the wheels on the cart."

He looked around for support from all of those listening. All the men agreed.

Stamford added, "We want to get the approval of the tried-and-true Green Power Network, a nonprofit, public service motivated corporation that's accepted by the big boys in power. They're for money and profit.

"The sand battery and the solar power from solar panels is new to a lot of people. I recognize that storage is a serious problem for the electrical industry and, if we can improve the situation in any way, we're making a great contribution. If we can provide four-hundred milligrams or sufficient micrograms of storage to keep our local substations working, it will be a tremendous contribution. It hasn't been done yet, but there's no reason why we shouldn't try."

Abbey responded, "Tom has learned about something new. A sand battery instead of the usual lithium batteries we have become familiar with."

Stamford replied, "Perhaps it may already be out of the box, but I'm betting on our location. If we can develop and sell the new concept to a

New England grid and maintain it in Cliffside, it will add a level of protection to the out-of-the-box element. Tom may know how to use this low-grade heated sand battery for solar panels, electric power, and perhaps—hopefully—even for wind power. It may help on my wind farm."

Tom interrupted their conjectures, "Have we covered it all here? Peter, do you and your group have enough to get started?"

Peter looked at his men and they nodded. He said, "I believe we do. We'll move on by getting in touch with the cleaning people and setting a starting date so they can be working while we're doing the studies and drawings."

Tom then continued, "We'll need an improved airflow within the building as well. Put that in your plans, along with receiving the solar power from our newly developed Agrivoltaic installation that we'll create on my farm. The solar power installation will provide the heat for the sand pit and fresh air for the roaring brook as well. Think out-of-the-box."

Peter replied, "Will do."

On the way out of the building, a policeman came up to Abbey. He pointed to the man at his side.

"This gentleman says he is the president of the Northwest Residents' Association. He says they had planned to purchase the coal plant to turn it into a school so children would not have to ride the bus for an hour each morning."

Abbey introduced himself and the others then said, "We have purchased the property for our new business development in this part of Cliffside."

The man said, "I'm Walter Kowalski. I used to work here twenty years ago, and I have always wished to see the plant redeveloped. Just what kind of business are you going to put in?"

Tom replied, "My family farm, the Carlton farm, has been your neighbor for years. It's to meet you."

Walter responded, "We've done business together for many years. I'm pleased to know we have local people purchasing the old coal plant. We were afraid it would be bought by outsiders."

Stamford introduced himself as Tom's uncle and said, "We plan to employ as many people from this area as possible and turn this plant into a productive commercial location."

He pointed to Peter. "The president of Wikiup Architects in Boston will be supervising the renovations. He is the man to see about future local employment. If you think anyone may be interested."

Walter said, "I'll spread the word around and call the farm to see if there is anyone looking for a job."

Tom answered, "Good idea. Tell people to call me to get in touch, and I will accommodate them if we can."

Walter shook hands with everyone. When he got to Peter said, "I'm glad Wickiup will cooperate with whomever the Northwest Residents' Association recommends for whatever jobs that become available."

Walter then left.

Abbey mused, thinking of Phoebe's high school teaching background.

"The building would make a good solar electric trade school for high school-age people who need on-the-job training."

Stamford said enthusiastically, "They could come to the wind farm to broaden their experiences as well. It would be a nice local use for the rooms in the entrance and on the second floor."

Tom added, "You guys are really something else. We'll build all of that into our curriculum. Instead of four offices for our employees, we'll have four classrooms. One could be on solar power, with the connection to the grid on the second floor in the back. One could be on dry sand heating energy, and a third one on hydroelectric production. The fourth one could be on the development of Agrivoltaic that we will be developing on the farm."

Peter added, matching their enthusiasm, "It will all be in the complete plans, and anybody interested in a job will be informed of it."

Abbey said, "Phoebe and I will get to work with the Board of Education to make it part of vocational training in Cliffside."

Tom added cheerily, "We've had quite a day. Let's eat and tell the ladies all about it."

He locked the gates behind them. Abbey walked over and thanked the police officers, who had been on hand throughout the morning. They then took off for the farm to have lunch.

The ladies had set the table with turkey, ham, and salami sandwiches, potato salad, and Boston baked beans. To drink, there was milk, lemonade, or beer. Dessert topped it all off, which was a chocolate-covered white cake.

Tom and Stamford told the ladies about the visit to the plant. It was met with great enthusiasm. The men dug into their food, letting nothing stand in their way.

Tom explained to Priscilla and his sister Nancy what they were going to do with the rooms in the front of the building. He explained that it was probably going to take at least a month once the cleaning companies got going. In the meantime, they'd be working on contracting twenty acres of Agrivoltaic panels fourteen feet off the ground and figuring out the plants that they were going to plant underneath.

Peter then said, "Wickiup will make four sets of the finished drawings. We will give one to the industrial cleaning company, another to the office cleaning company, and two full sets to Tom and your board of directors."

Abbey added, "It would be a good idea if we had selected drawings for the city's Board of Education and the Police Department. Will that be a problem for you, Peter?"

Peter responded, "Not at all, Abbey. As a matter of fact, I'm glad you thought about it, because it probably wouldn't have entered my mind.

I'll be glad to provide anything that the Board of Education or the Police Department would like. If they need anything specifically done, I know the security force is going to be important to operating smoothly. Tell them to just give me a call."

Tom added, "Mom, we must keep in touch with Walter Kowalski. He is the head of the local civic association."

She added, "I'll do it right away, Tom."

After everyone had their fill of lunch, Tom said, "I know Stamford, that you have to get back to sea tomorrow, so if it's okay with Abbey and Phoebe, I'd like to contact the solar panel suppliers right away—I think they called themselves Morning Sun—and got them started on our twenty-acre installation. They'll probably want a down payment. If that's okay with you, then we'll need a check for a deposit from Forever Blue for the company to get started. Any questions?"

Stamford replied, "If it's okay with Phoebe and Abbey, it's okay with me."

Abbey replied, "No problem here."

Phoebe added, "Okay with me."

Tom then went on to say, "In the two-hundred-and-fifty years that the Carlton Farm has been in Cliffside, our farm has produced vegetables, dairy products, and large quantities of hay. Our products have changed with the city's requirements, and now they're about to change again.

"This time it will be a cash-crop solar-panel installation we're creating. Solar panels provide electricity, not just for our home or the barns, but for general needs of Cliffside. The developer in Boston focuses on large solar installations and battery storage. This allows an excess of electricity to be fed to the power grid.

"This company, Morning Sun, is who I hope to employ to do our Agri-voltaic installation. Solar panels will share space with crops so they both can survive and grow on the twenty acres or more of the Carlton farm plateau located north and west of Cliffside. The excess solar power will then be directly transmitted to the grid or used to energize our sand pit and stored for future use.

"The technology of Agrivoltaic involves adjusting the height of the solar panels to about fourteen feet above the ground as well as adjustments for sharing the light between them and the plants below. The panels will be high enough to accommodate any farm equipment, the workers working the crops directly below them, or grazing animals. The spacing and the angles of the panels allow light to reach the plants below and has the added benefit of shielding them from intensive heat loss. The electricity generated gets loaded to the grid through our former coal plant, and the power will find its way to local use.

"The solar panel firm from Boston will design and construct the installation and provide supervision on a contract basis. Carlton Enterprises will own the establishment and the power it produces as well as any financial or miscellaneous benefit. Carlton Enterprises Inc. will lease the land from the Carlton family farm."

Tom told them about the project in Arizona, where he saw a threefold increase in crop yields when under the solar panel system and up to a fifty percent reduction in irrigation requirements because the panels provided shade as well. The Arizona installation under panels released water into the air that cooled molecules, creating a beneficial symbiotic relationship between the plants and the panels. He explained his hope that the same would also happen in Cliffside, even though the climate was different.

"A farm in Maine produces four-point-two megawatts of power, and it has been estimated to produce 5,468 megawatts annually, roughly the amount to power about five-hundred homes. The company has also been fortunate to have developed a commitment to the social environment to achieve those goals. As a result of the efforts around the country, the use of Agrivoltaic is growing all over the United States. The United States Department of Energy office said their 2021 report that fourteen gigawatts of power were generated in dual-use systems like Agrivoltaic, which is roughly the equivalent amount of electricity necessary for two million homes in the United States."

Tom cautioned, "There is a negative side. There have been lawsuits by some people who do not want the installations in their backyard. They have obstructed the developments in their towns in some areas. Fortunately, we have cooperative people in the northwest part of Cliffside, so I don't anticipate any serious NIMBY problems or violence toward our factory or the installations. However, we will be prepared with a security force if that becomes necessary.

"Ultimately, though, according to the Agrivoltaic learning lab at the University of Arizona, it's the taste of the veggies and their appearance that's critical. Potatoes and Brazilian squash are preferred in these conditions, but other vegetables have been trialed successfully. All farmers know that rows of vegetables in the sun always taste better, and the farmstand customers that we have will soon know the benefits of the symbiotic relationship we're developing.

"It is my intention to have our factory and our twenty or more acres of Agrivoltaic open to the public to encourage their support. But we'll maintain an active espionage and security force."

Peter added, "We had a wonderful day and a marvelous meal. We plan to be here on a regular basis while the cleaning is going on, but we'll say goodbye for now and give you a call when we're ready to come back again."

They shook hands all around, and Peter headed back to Boston.

CHAPTER FOUR

Martha, Tom's longtime girlfriend, was in her fourth year at the local state university, and she agreed to Tom's request to research Arizona, Maine, and Massachusetts's Agrivoltaic experiences to give the business some background information while they were putting the enterprise together.

Nancy approached Tom and said, "We have to talk to Mom; she seems quite concerned. She's lost in your enthusiasm over the Agrivoltaic and the financing. Don't you remember how Dad used to tell us, "Give her time to warm up and then she'll follow along."

Tom sat down with her and said, "You're right—I'm moving ahead too fast with buying the coal plant and organizing the corporation suddenly, from nothing. Now, we are talking about taking twenty or forty acres of our farm into the enterprise. I can see why she would be concerned about the family farm and its financial condition. I'll slow down, and we will talk to her privately. Shall we do it right now?"

Nancy nodded.

Tom invited his mother to sit down next to him and said, "I just wanted to take the time to tell you what I'm trying to do. Agrivoltaic is a new, different, and better way to grow our food by controlling the heat and the water supply that the plants require. Farmers and our workers will notice cooler conditions while working under the panels. Farms around the country are being funded by the National Science Foundation, and they are anxious to have more local information about different developments in various parts of the country. I hope we can be one of the contributors to their search for information.

"My previous studies have shown that partial shade on plants like tomatoes, peppers, chard, kale, sweet potatoes, cabbage, and onions improve the product yield considerably. By having the solar panels above them, we'll be created better growing conditions. The simple solution is growing the veggies and letting the solar panels trap the sun's heat. In cooler climates and where there is water, we can protect the plants in the areas where water is diminished by excessive heat evaporation. That helps. Cooler panels convert more of the sun's energy to usable electricity. The international Bloomberg News service reports, *'The year 2022 was a blockbuster year for Energy Trends. Nowhere is it spreading as rapidly, dramatically or more successfully, than in the nations that have built large solar panel installations.'*

"There is even talk about creating a national electrical nuclear utility. Many are already government sanctioned monopolies at city levels, so the principle of a national monopoly and energy is already established. The intermittency of wind and solar means it might make sense to run new long-distance power lines between cities.

"I think that our new twenty-acre-or-more solar farm will actually be a resource in this new energy era. And as Dad used to say, I want to 'let it run'."

Nancy interjected, "Mom, I know all of this is new and confusing, and there are a whole bunch of guys, trucks, and equipment on our farm, but we can set it up so it doesn't interfere with our customers coming to the farmstand, and it might even bring in more people and better vegetables."

Tom added, "We're in control of the corporation—we own it. Forever Blue Public Trust has pledged its financial support, so I think we're all right. Stamford agrees as well."

Stamford, who had been listening, came up quietly. He said, "Yes Priscilla, Tom has been very careful, and it will not threaten your farm or your business in any way. It will not only give you better products, but it will improve your financial condition by providing you with a monthly check."

Priscilla said, "I know you all think well of the new Agrivoltaic installation; I'm just a little apprehensive about it all. But I have great confidence in you, and I will give it my full support."

The next morning, Tom was on the phone with the Morning Sun Corporation. He explained to them his desire to purchase a twenty-acre solar installation of Agrivoltaic. They agreed to look at the area to see if they could accommodate his planned idea and set up a date for a week later.

Abbey and Phoebe were on their way to set up the factory security system with the police and then visit their former Board of Education boss. Now, they were talking to city councilman, Stanley Johnson, about electrical vocational education at the newly renovated coal plant—now being called the Green Power Network Facility.

Peter Whitehorse and the architects at Wikiup were busy finding the proper corporate cleaning crews for the coal factory and the officers who would help protect it.

Tom spent the week at the north end of the farm with some of the employees to clean out old brush and cut away the growth to create an access road, including a couple small trees at the end of the road. The lumber would be used by the construction people.

On the appointed day, a small van with solar panels painted on the sides pulled up in the farm store parking lot. Two men and a woman got out of the van and walked into the store.

Tom greeted them, introducing himself. The man introduced himself as Roger Donovan and his wife as Mary Donovan, the owners of Morning Sun. The second man, George Rolston, was their electrical engineer construction supervisor. They shook hands all around.

Tom said to George, "I saw you at a seminar on agriculture and Agrivoltaic energy three years ago at the university I attended. I was impressed, and I graduated with my master's degree in electrical engineering last year."

George answered, "I thought you looked familiar, congratulations! I'll enjoy working with you on this new project."

Tom said, "Come in—I've got the other room set up with a table and chairs and we'll have time to talk."

They all walked into the small room next door that was set up with coffee and pastries. Mary said, "This is very welcoming, thank you very much. It was a long ride from Boston."

Tom said, "Thank you, it's my mother's touch. We've owned this farm of six-hundred and forty acres for five generations."

There was a group of pictures on the table of different views of the construction from start to the finish.

Tom said, "Just a moment."

He left, and they heard him call for his mother and Nancy to come in and see the pictures. They came in quickly. Tom took them over to the table.

Tom laughed, saying to their guests, "You can see who's the boss on this farm."

Mary, Roger, and George all laughed. George then said, "Yes, she's a good businesswoman, no doubt about it."

Priscilla answered, "I hold it down with my youthful enthusiasm—I'm afraid I've been running this business for thirty years now."

Tom continued, "My studies have been primarily on Agrivoltaic in several geographic regions that experience food production challenges and water shortages but that generally have an overabundance of sun energy.

"I want to build a study for the New England area of the United States, similar to those that have been started in Arizona and Maine. I don't believe they have reinforced studies to seriously track the grid history. In addition, our state, like many in the east, has serious housing shortages and serious food shortages due to drought.

"I think food shortages and housing shortages seem to go together. By providing additional food from a newly developed green power, we can make a marginal difference on the availability of reasonably priced food."

Priscilla Carlton looked at the three of them and said, "I know you all think well of the new Agrivoltaic installation. I'm still a little apprehensive about it, but I have great confidence in my son. I'll give him my full support."

Roger said, "If we are your supplier, the pictures are yours to keep, along with the new photos we will be taking during construction. As you can see, the plants are growing very well underneath."

Roger then said politely, "Okay, let's go down to the proposed location and see what has to be done."

Tom answered, "Okay, I have my truck out front—just follow me."

They all climbed into their vehicles and drove down the road about a thousand yards. Tom turned onto his newly cleared road through the grass, up to the twelve-foot bank, and onto the plateau. They followed him closely in their truck and pulled alongside. He came to a halt.

George asked, "Is this whole area part of your farm?"

Tom answered, "Yeah, for more than a hundred years."

Roger exclaimed, "A forty-acre Agrivoltaic field against that mountain background will be a New England showcase."

Tom said, "Hold on, I'm only talking twenty acres for now."

Roger answered, "I'll give you the additional twenty acres at my cost. We'll make it the finest installation in New England."

Tom lowered his voice and said, "During the farm's hundred-and-fifty-year history, it has produced vegetables, dairy products, and hay. Now, the farm's evolution has a chance to join a changing market and fight against climate change with a new kind of cash crop. I have seen it moving in that direction, but like our forefathers, we must jump in with both feet. Having forty acres next to this metropolitan area will give us a major advantage for green electrical power as well as for food, and we must not forget any additional family income.

"But it's a major investment to have this symbiotic relationship between plants and panels. Now, you're offering your adjustable panels fourteen feet above the ground that can be moved, as need be, to capture the power of electrical sunlight. As a farmer, everything depends on the success, taste, and appearance of the produce."

Roger said, "I have learned the more mobile our panels are, the better the potatoes, squash, blueberries, tomatoes, and other crops taste. The power yield also increases. A new installation in the state of Maine produces, they tell me, forty-two megawatts of power now, and they're estimated to produce 5,468 megawatts in a year. The famous Knowlton farm in Grafton, Mass, with eighteen-hundred solar panels also provides a steady winter income and renewable electricity power for twelve-hundred homes."

George added, "I wouldn't be surprised if you get your money back in agricultural products and sellable power in five years."

Tom thought for a long moment as they walked around the proposed area for the array. Then he said, "Roger, I'm convinced enough to carry your forty-acre proposal to my board of directors. I'll call you in the next day or so and let you know whether it's going to be twenty or forty acres, but it will be one of them."

Roger said, "Okay, we got a deal. I'll prepare our people for the installation."

Then he said to George, "You are now their supervising electrical engineer. Take some photographs of the area, ask a few more questions, then we'll get back in the truck and leave for Boston."

Tom immediately went to the telephone, called Stamford, Abbey, Phoebe, and his mom and told them about his meeting with Roger from the Morning Sun, George, and their construction crewmen.

Abbey asked a few questions, but Stamford said, "I'll quote your dad Tom: 'Give the horse his head and let him run with it.'"

He continued by saying, "I'll call the Forever Blue sustaining memberships and invite them to join us. I think it'll be okay."

Abbey added, "It will increase the enterprise's reliability, and later, if the grid buys some of its power, that will help, too. You'll also benefit Tom, from the farm vegetables income, which should be increased as a result of this product."

Tom called Roger Donovan back and told him to go ahead with the forty acres of construction at the far end of the plateau. Roger thanked him very much and hung up the phone.

CHAPTER FIVE

It took ten days for Morning Sun's van, followed closely by a tractor trailer, to show up at the farm parking lot. They arrived at eight in the morning.

Tom was happy to go out and greet them, but he was surprised to see the tractor trailer loaded with a large quantity of woven wire fencing, four men, and a six by twelve-foot sign picturing their name, address, and a beautifully colored picture of the fourteen-foot Agrivoltaic panel above healthy growing vegetables.

Tom approached George, who was driving the van. He said, "Good morning, I didn't expect you to arrive so soon."

George answered, "Roger put your forty acres at the top of the list and said again, 'We'll make it the best array in New England.'"

Then George continued, "I brought enough fencing to go around three-fourths of the plateau, a gateway for your new access road, and some additional fencing to protect your farm store parking lot."

Tom asked, "Will we need fencing?"

George answered, "The boss got tired of losing tools and other miscellaneous items to onlookers and invested in a thousand yards of fencing. After that, the theft stopped. We also put a security guard on duty for twenty-four hours; I hope you don't mind."

Tom answered, "You know what's best. You have the better part of a million dollars invested in manpower and materials. "

Tom then asked, "Can I help in any way?"

George said, "I have the men and all the equipment I need. And my office trailer is on its way, so we're ready to start fencing. If you have a large tractor I can borrow, I'd appreciate the use of it for a while."

Tom answered, "Yes, we have a large John Deere and a driver at your disposal anytime you want it."

George then added, "I'd be pleased if you'd have him drive over the forty acres that we're going to use several times so we have a nice flat area for our survey while my workers put up the fence on three sides and the gateway at the driveway. They'll need your advice on the fourth side that's next to your farm store and parking lot. They'll probably be on that fourth side by the end of the day."

Tom said, "I'll put a man on the tractor right away and hurry him toward the barn."

George drove up the access road with the van and the truck until they arrived onto the plateau. They parked and got out, and one of the men shouted, "Wow, this is a super sight. I can see why Roger is so impressed."

George looked at him then replied, "The Carlton family has owned these six-hundred-and-forty acres for over a hundred-fifty years."

Another of the men said, "They must be rich."

George said quietly, "Only in land and, of course, the farming business. That's why they're reaching out for a new source of winter income from our solar power Agrivoltaic arrays. Roger hopes it'll be as good as the big one in Grafton. I think the mountain in the background was what really caught his attention. Tom hopes that we'll make it happen for the whole Carlton family."

At the same time, Priscilla came back with Tom and the John Deere. She overheard the man talking about the site with pride. She said to Tom, "Tell them some of our old family stories about the location."

Tom said, "Okay, Mom."

But George interrupted and said, "Thank you for the offer to tell us some of your stories, but we don't have time today. I'll be nervous until we get our fence all around."

Tom said, "Okay, I'll set up a campfire and will tell you a story while you guys are sitting having lunch. How about that?"

George smiled and said, "Okay, Tom, let's do that."

He turned to his men and said to those setting up the field kitchen, "You help Tom, and the rest of you, let's get going on the fencing."

The work went quickly. First, two men surveyed the boundaries. Two men drove wooden stakes in the ground every eight feet. Then a truck pulled up with one man walking alongside and another in the bed next to a pile of long, steel posts. The truck stopped by each wooden stake, and the man on the ground took one steel post from the bed of the truck. He held the steel up beside the wooden stake, and the man on the truck took a sledgehammer and drove the posts three feet into the ground.

Two additional men came along with a large rolling device that held an upright bale of woven fence wire that they unwound and attached to each steel post.

Tom watched, fascinated by their skill and efficiency. He said to George, "That's quite an efficient arrangement."

George said, "Our boss, Roger, designed the equipment himself. We don't have any more problems with equipment loss."

Tom went back to making the campfire. At lunchtime, the men

gathered around the fire while eating their lunches. At this time, Tom started his story.

"Josiah Carlton and his two sons were mountain men in the 1600s in this part of New England. They were guides and hunters for the British Provincial Army that controlled the area. During the course of their duties, they discovered an unusual forest on a perfectly flat plane and were fascinated by it. The two sons suggested they acquire the property and turn it into a farm. Josiah approached the Colonel in charge of the troops and suggested that he and his two sons could work as British special soldiers for three years. In return, they would receive a fourth of the wooded property they'd described.

"The British Colonel accepted the deal. Josiah and his two sons worked diligently for the British Army for three years. True to their deal, the colonel presented Josiah with a Royal Land Grant for six-hundred-and-forty forested acres.

"To the west, about five miles out, is a wild mountain stream. It was—and is—so wild and so swift that the locals call it Cataract. It drops down into a wide flood plain, creating the river on the south and a wide damp area to the north that is dry in summer and wet the rest of the year."

Tom continued, "As a matter of fact, the river continues down to the bay. It provides Cliffside and Bayside with high ground and dry communities."

George didn't want to miss the story, and he listened intently. Now that it was over, he said, "Alright, guys, back to the fencing. Let's get busy unloading and get to work. Our office trailer will be here directly with two more men to help with the fencing to get it done by the end of the day."

Meanwhile, Abbey and Phoebe enjoyed breakfast at the Painted Lady.

Phoebe commented, "I hope Tom is successful in getting that solar panel company to come out here from Boston. It's a pretty big hike."

Abbey responded, "It's not too bad these days with all the superhighways. They'll get them up into Northern Maine in a few hours. Though, I had a salesman tell me one time that it was a shorter distance between Bangor and New York City than from Bangor to Presque Isle on the Canadian border. Of course, that was before President Eisenhower built all our new superhighways. Our problem is really going to be getting our chief of police to give us sufficient manpower or security at our solar plant."

Phoebe added, "And our new councilman, Stanley Johnson, my former boss at the Department of Education, is buying our vocational training idea at the coal plant after it is renovated."

Abbey said, "I'll call the chief, Ron Larkin, for an appointment. Can you call Stanley for an appointment this afternoon?"

Phoebe said, "Okay. You set up for a ten-thirty appointment with the chief, and I'll set up a one-thirty appointment with Steven. We'll get lunch at the diner in between."

Abbey smiled and nodded.

"Agreed. I'm in the mood for a good fish sandwich and fries myself."

They went their separate ways to their apartments in order to make the calls. At Phoebe's door, she said to Abbey, "I'll meet you on the porch at ten a.m. if it's a go with the chief."

Abbey called back to her, "I'll be there."

Abbey called the police station and ask Ed for the chief. He answered.

"Good morning, Abbey, I haven't heard from you lately. What have you been up to?"

Abbey answered, "Are you familiar with the Carlton farm on the plateau in the northwest part of Cliffside?"

The chief said, "Sure. I buy from them all the time. I understand Old Tom died last winter."

"That's true," Abbey replied. "His son, Tom, just graduated with a master's degree in electrical engineering from Rensselaer University and has stepped into his father's shoes. He wants to add Agrivoltaic solar power to forty acres to supplement their income during the wintertime."

The chief exclaimed, "He must not only be smart but aggressive."

Abbey answered, "Three years as captain of the college lacrosse team would tell you that. He gave the board members of Forever Blue Public Trust a pitch for the investment at his birthday party. Stamford Pettibone, we discovered, is his uncle on his mother's side, and Priscilla, Tom's mother, is a longtime friend of Phoebe's.

"Tom wanted Forever Blue Public Trust to back the purchase of the old coal electric plant. He intends to convert the plant into his solar power facility. Then he wants to connect the forty-acre solar array with his coal plant to establish a power distribution system. Hopefully, in cooperation with the Cliffside Bayside Power Distribution Substation.

"Then his new Green Power Network will be invited to participate with the New England grid, just as Stamford does with his eight-tower wind farm fifty-eight miles off the North Atlantic coast."

The chief whistled and said, "That youngster really thinks out-of-the-box."

Abbey answered, "You're going to say that again. Stamford and several Forever Blue members are backing it on behalf of Tom Carlton's Green Power Network expansion and climate change efforts. They even went so far as to advance his new corporation, Carlton Farm Enterprises Incorporated, at a hundred and ten thousand to purchase the plant. Raphael Sullivan, has registered Tom Carlton, Stamford, Mrs. Carlton, Nancy, Phoebe, and I as board members."

The chief said, "I'm getting the feeling you're going to ask for something."

Abbey laughed and said, "You're right. We'll need twenty-four-hour security against ordinary perpetrators and protection from burglary and vandalism."

The chief took a moment and then said, "I can give you a drive by each hour with a coffee visit every three hours for seven days. Inside, I can provide an auxiliary policeman from three to eleven and eleven to seven, plus twenty-four hours on weekends and holidays."

Abbey responded, "Thanks, Chief. If Forever Blue or the enterprise corporation can help the police in any way, just call me."

The chief answered, "Thanks, Abbey. At the moment, we're enjoying our sealed bodycams, and our police union is happy with our new public reputation."

Abbey thanked him again and left his office. Once in his car, he phoned Phoebe and asked if she'd arranged a meeting at the City Council. She said, "We're all set for one-thirty. You can pick me up for lunch now if you're free."

Abbey replied, "I'll be right there."

They arrived at the diner a little early for the lunch rush, so they had their choice of booths. A voice called to them from the large round table in the rear. It was detective Steve Corcoran. Abbey waved, and they went back to join him.

They greeted each other, and Phoebe sat down beside Steve with Abbey opposite him. Steve opened the conversation.

"The chief brought me up to date on your new adventure into solar power and the old coal plant backed by Forever Blue, thanks to Stamford's influence. I didn't know he was related to the Carlton farm family. You guys sure do get around."

Phoebe laughed and said, "We are busy bodies at the Painted Lady."

Steve chuckled and said, "Abbey, do you know that Joe Hardcastle? Who owns the Hendricks Hardware Store, the home fuel oil business, as well as the GMC truck dealership. He had his eyes on that old coal plant before your swift move with Hawthorne Realty frustrated his ambitions. He isn't happy about it. He also has a fortune in fossil fuels and is on the Boston Fossil Fuel Dealers Board."

Steve added, "Just for fun, Hardcastle's mom is on the City Council. She'll try, I'm sure, to hold up your reactivation of the commercial zoning and grid permits for the coal plant."

Phoebe spoke up. "We're going to see our new Councilman, Stanley Johnson, at one-thirty today, and we'll alert him. He and Carmen will be prepared to help."

Steve replied, "After your sizable contributions through campaigns by the Forever Blue Public Trust, they're obliged to support the Carlton family. Besides, I think the local people, through association, would rather have solar power than fossil fuels—they should be on your side as well."

Abbey offered, "we've already met Mr. Kowalski, of the Residents' Association, and he seemed quite friendly."

Steve responded, "Strengthen that local connection; they're pretty active. I've met a few of them over the years. I'll call some people and show them that the police favor the new Carlton Farm Enterprise."

They ordered lunch, and Phoebe told him about the vocational plans for the eight vacant office rooms in the front of the building. Four on the first floor, and four on the second floor. Steve thought it was a fine idea; the city of Cliffside and Bayside need good vocational education.

Then she went on to say, "The solar panels, hydraulics, hydrogen, and all sources of power should help, and the local Power Distribution will help the city be more prepared for emergencies."

Steve added, "I'll talk to Harvey, the lawyer for our Police Commission, and see if he can't get the mayor to back your program."

They thanked Steve for his support and invited him to visit the plant once it was all cleaned up.

After lunch, Phoebe and Abbey drove over to the City Hall and the councilman's office. They were right on time, and Stanley was waiting for them. He led them into his private office and offered them a seat. Then he asked, "How can I help you?"

Phoebe answered, "We need your support with our new public project—the Green Power Network—which is in our new purchase: the old coal plant."

Stanley sat back in his chair and asked, "Tell me about your new project."

They went into the whole story. From Tom Carlton's graduation from college to their purchase with the help of the Forever Blue Public Trust to the old coal plant and the creation of the Green Power Network Inc.

Abbey added, "Tom's purchasing Agrivoltaic solar panels and installing them on forty acres of his farm plateau. It is also designed to create additional electric power to assist the community in case of emergency. Therefore, it is incorporated as a nonprofit public service corp."

Phoebe continued, "Tom's mother, Priscilla, and her foster brother, Stamford Pettibone, who owns an eight-tower wind farm on the Atlantic Ocean, has joined the cooperation and is contributing to their surplus power as well."

Stanley interjected, "Boy, that young fellow believes in hitting the ground running, doesn't he?"

Abbey added, "He will connect the coal plant as his power unit to use the surplus power from the forty-acre Agrivoltaic array to assist public projects. We therefore need licenses from the city of Cliffside to connect to the grid and the rights from the city to be a nonprofit public service corporation. Stanley, that's where you come in."

Abbey continued, "We hired, through an architectural firm called Wickiup from Boston, two firms: one to clean up the manufacturing coal plant and another to clean the office part of the facility. We have arranged with the Police Department to have round-the-clock security. Our attorney, Raphael Sullivan, is handling all the details."

Stanley asked, "Just what is it you want me to do?"

Phoebe answered, "The coal plant has four unusually large offices on the first floor in the front wing of the factory and four on the second. I'm considering turning those offices into vocational classrooms on public and electric power, with the city's blessing and support."

Stanley smiled. "My favorite high school teacher has never lost her love for your children. God bless you, Phoebe. But the city Board of Education and the City Council may think you have lost your mind."

Phoebe insisted, "We can train electricians, power company lineman, car builders, and house owners who are putting electricity into their buildings; they will all have the benefit of the new green power electricity over carbonized fossil fuels."

Abbey added, "It's thinking out-of-the-box. A small city, training its people to use non-fossil fuel power to protect itself in case of emergency. A real step in climate change control."

Stanley responded, "I'm with you, and I know Carmen will be, but it's going to be a hell of a climb for the budgetary City Council. They might go along if you were self-supporting; they might give their okay then. Let's run the flag up and see who salutes it. I'll get back to you, just give me a couple of days. Leave me your phone number."

Abbey added, "If money is the primary problem, tell them to relax. It is Tom's intention to be totally self-supporting with the surplus power that he'll have from his forty acres of solar energy that will be distributed through the former coal plant."

Tom and Phoebe got to their feet. She gave Stanley a hug and said, "Thank you very much for listening. We have a great idea, but it takes a couple of experienced politicians to get it to slide through the City Council."

They were surprised to see all the people gathered at the plant. The architects were providing coffee and danishes for everyone from ten-thirty to three every day. When they drove by, Abbey stopped the car and said, "Let's go look. The people seem to be happy with what they see."

One of the supervising policemen, Jim, who was driving the cruiser, said, "The industry cleaning service's boss hired several local people to carry out portable machines, equipment, office furniture, and office supplies, and put them in the trailer. The idea is to lock the trailer to protect them."

Phoebe asked, "How many did he hire?"

Jim the policeman answered, "Mr. Kowalski supplied twenty men and four women. Peter Whitehorse has their names, addresses, and phone numbers. He and the industry cleaning boss agreed on their salaries, which the architects will pay and put on your bill, okay? We supervised the whole activity, and they're going to give an honest day's work for an honest day's pay."

Abbey replied, "Very good officer, thank you very much."

Mr. Kowalski, the president of the Northwest Residents' Association, came up as they were talking. He seemed relieved when Abbey and Phoebe approved what they had done. "I assumed you wouldn't object. I added my name to the Association's list as a personal representative."

Abbey said, "I will personally send you, as the personal representative of the Residents' Association, a check for a thousand dollars for the Association's close cooperation."

Kowalski tipped his hat and said, "Thank you very much."

Jim, the police officer, said, "Peter Whitehorse had his architects set up offices on the first floor in the first room on the right."

Mr. Kowalski offered his help on a regular basis.

He told Abbey, "I have a friend in the association who is a retired foreman in electrical. His name is Henry Lipman, and he worked for the Cliffside Power and Light Company before he retired. Since you're going to be looking for help with that kind of experience, I thought you might want to talk to him. Maybe he could be of some help to you. He lives just half a mile down the road. I told Peter Whitehouse about him, and he said Tom Carlton is the owner and the electrical engineer, so he would like to meet this man and test his technical ability."

Abbey said, "Peter's answer is absolutely correct. I'll also mention that to Tom when I see him to make sure he knows about Mr. Henry Lipman."

Kowalski said, "Thank you very much. I'll tell Henry, so he'll be prepared."

He shook Abbey's hand and left.

Abbey said quietly to Phoebe, "Looks like we're going to be lucky and get help from this neighborhood."

Jim's police companion said to Abbey, "When you see Peter White-house, tell him thanks for the coffee break, but we've got to finish our rounds."

Abbey agreed and gave him his number, saying, "Thank you for your help. The forty-acre, Agrivoltaic, fourteen-foot array is being constructed as we speak. The coal plant is undergoing a total cleaning of all fossil fuel debris. We should be ready to occupy it in about thirty days."

Jim started the cruiser, and they drove off.

CHAPTER SIX

Abbey took Phoebe by the arm. "Let's take a look at Peter's new office." They walked up the front steps of the plant and entered the building and were confronted with work men and women with white overalls. They turned right and entered an office with big double doors, closing the hustle and bustle behind them. These large offices in the front wing of the plant were for the company offices that dealt with the public that served in acquiring supplies.

The other office across the hall had large double glass doors to make access to desks and other large furniture possible. The transmission and other communications with the local power distributor were in there, too, and if necessary, the New England Grid and the National Power Grid.

They were surprised to see a man sitting in front of a freestanding, six-foot high, three-by-five-foot monitor. They watched as the man turned and recorded numerical dimensions into three-dimensional full-screen drawings. Abbey said, "Our architects certainly work magic."

Phoebe looked in at the office and added, "It's as large as my high school classroom."

Abbey smiled knowingly and said, "Yes, it is."

Peter, who had come in, asked, "Do you like our new style monitor and our exclusively developed software? Meet Harry, our computer guru, draftsman, and artist."

Peter turned to Harry and said, "Say hello to our employers and friends, Phoebe and Abbey."

Harry turned, got up, and shook hands with them. He then returned to his creative task.

Peter said quietly, "He is not only an artist, but a top-notch computer man. He'll start turning the computer images into reality. From there, we'll need key employees who know what they're doing with steam, turbines, generators, and the solar power coming from the forty acres of Agrivoltaic on Tom's farm. Newly employed experts who we'll train will train the new employees. Your Director of Security will have to ensure their confidential know-how and the daily running of the plant.

"Those of us at Wickiup will be with you all the way. If Tom needs our skills for his installation at the farm just let us know we'll be there to help."

Abbey said, "I'll pass that word on to Tom."

Phoebe added, "I'll talk to Pris and Nancy to get the ladies' angle on it."

Then she suggested to Abbey, "Let's you and I visit the farm now, and get the Carltons up to date."

Abbey agreed, and they returned to his car and drove north to the farm's plateau. On arriving at the farm's parking lot, Abbey spotted Tom off to the west near the construction of the solar panels, and he gave him a blast on the horn. Tom turned toward the sound, recognized Abbie's car, and waved as he came toward them.

Phoebe then said, "I'm going to visit the ladies in the house, and bring them up to date to get their opinions."

Abbey agreed and started walking towards Tom along a new truck track inside the new fencing. When he reached him, they settled down together in the grass.

Abbey told him about his visit with the police chief after picking up Phoebe and having lunch with Stanley at the City Council. Tom was pleased with their willingness to support the new corporate effort.

Abbey reassured him by saying, "During our visit, we stopped at the factory. We were impressed with the two cleaning companies and Peter Whitehorse's supervision. Peter supervised hiring twenty men to work in the factory and four women to work in the office part of the plant."

He smiled and said, "Tom, wait till you see Peter's work office on the first floor. They have a large computer and a self-standing, six-by-five color monitor."

Tom said, "I've seen one of those in school when they gave a demonstration. That equipment certainly can be impressive in the right hands."

Abbey replied, "They have an artist and a computer operator, a Native American from Algonquin University named Harry who is turning out all of these drawings. He'll be producing the working drawings for all of the men we hire and will finish drawings for management at the appropriate time."

Abbey added, "Mr. Kowalski has a friend in the association who was a foreman for Cliffside Light and Power before he retired three years ago. His name is Henry Lipman, who lives nearby. You might want to hire him as a manager—he has a fine electrical background, according to Mr. Kowalski."

Tom answered, "I have Kowalski's number. I'll set up a meeting with him and Henry Lipman."

Abbey responded, "Good, now tell me about these forty-five acres of upright steel columns for solar panel power. I thought it would just for forty."

Tom answered, "The guy to talk to you all about it is George Ralston, the man in charge of installation and maintenance."

Tom turned and whistled without getting up. A tall young man turned over by the installation with a red hat on. Tom waved him over.

When he came up, he asked, "What's up Tom?"

Tom responded, "Have a seat and shake hands with Abediah—we call him Abbey. He's the retired chief of detectives and the treasurer of Forever Blue Public Trust, our five million-dollar financial backers."

George reached over and shook Abbey's hand as he sat cross-legged in the grass.

He said, "It's a pleasure to meet you, Abbey."

Tom added, "Abbey needs a non-technical introduction to the installation, and wants to know why it's going to consume forty-five acres instead of forty."

George thought for a moment.

Then said, "I'm sure you're aware the sun heats the world. What you may not be aware of is that the heat is so intense that it could easily destroy all life on this planet if we didn't have the atmosphere around the Earth. That protective blanket is being thinned by an increase of carbon in our air. The intense carbonization of the atmosphere, caused by the burning of fossil fuels to give us electric power, is thinning the protective blanket. We must stop increasing carbonization from fossil fuels and find a new, safer way to provide the power we need for our existence. It's a matter of protecting the atmospheric blanket of the world.

We have learned we can get power from our hydroelectric dams. We can also process hydrogen from the air and process solar energy to gain the power that we use each day. We have been successful in getting energy from solar panels. All of these methods are what we call green energy. We can, through this process, create power without putting any carbon into the atmosphere. It's expensive, but it's creating the conversion we need to survive. Some of our country, and parts of the world, are already suffering damages from droughts and floods. This is caused by the thin spots in our covering blanket already."

Abbey said, "I realize you're oversimplifying, but the metaphor of the blanket encompassing the world while we sit in grass is very effective."

They both laughed at his reaction to the blanket metaphor.

George continued and said, "Our construction is using steel to support the solar panels above. We will sink steel rails four feet into the ground, surrounded with concrete at their base. The above ground rail is exactly fourteen feet high, and we'll set them in teen straight rows, four acres long, fifteen feet from each other. The math leads to forty-five acres of steel rows.

"Next, we'll attach solar panels to the top of the steel structure. Then we'll have forty-five acres of steel columns with solar panels on top. We're going to add another three acres for the body of the construction, where the master control solar panels will be located at the top of each of the rails.

"It is estimated that the total forty-five-acre array will produce six megawatts of power for a total of one million megawatts. In the beginning three acres, we will install a transformer to receive the power from the array and then transfer the power into your plant, as well as the sand pit."

Abbey asked, "How does the power get from the transformer to the pit?"

George answered, "It will be transferred underground from the transformer at the installation through an underground pipeline and attached to the receiving equipment in the building. I could give you a lot of technical language that gives you the specifics of that transfer of power, but the real expert will have to do all of that."

Tom added, "Our corporate attorney will arrange for the rights to receive the power lines that run underground—between the transformer, to the receiving equipment in the plant—and our new employees with the technical understanding and experience will have to take it from there to distribute it through the plant. I've already talked to Raphael Sullivan, and he's on the job right now."

Suddenly, Abbey's phone started ringing. He answered it. It was the Police Lieutenant.

He listened briefly then said, "I'm with Tom and Roger right now."

Tom said, "I want them to hear this."

Abbey said, "Lieutenant, please repeat what detective Jim Sweeney said."

The lieutenant repeated himself: "You are being, and have been, observed and photographed by a drone. I've taken their ID and will investigate, but I think you should know because it's heading back your way."

Right after he said it, a drone appeared above the farm and moved toward the installation of the solar array. It hovered at about a thousand feet in the air above the platform, obviously taking pictures.

Abbey said to the detective, "Tell the chief to put out an arrest warrant for the violation of our constitutional right of privacy against the owner of the drone and their employer."

The detective responded, "Can do—out."

George said, "Somebody is curious about what we're doing, both at the plant and here at the farm. I'm going to call my boss and let him know what's going on. He may want to do something about it himself."

Then he asked Abbey, "You're a local businessperson—who could be curious or upset by what we are doing here on the farm and in the old coal plant? Anyone interested or annoyed enough to be a threat?"

Tom answered, "I haven't given it any thought, to tell you the truth.

They must be fossil fuel people. Who else would feel threatened by a large green solar power installation?"

Abbey offered, "The only one I can think of is McCormick's trucking. They deliver diesel oil, propane, and natural gas. It is a very large operation. Come to think of it, McCormick is providing EMT trucks to the Cliffside medical ambulance units. They are looking to dominate the business both here and in Bayside. They're close to Glenside Insurance, an ultra-conservative political group. We have had encounters with them several times in the past. If McCormick is behind the drone, we'll know soon, and we should prepare for his political opposition. The New England fossil fuel supporters of Boston may be behind the move."

George said, "I'll tell my boss so that he will be prepared with tighter security around the installation."

Abbey and Tom both agreed to that.

When Tom got back to the office, they had a call from Kowalski suggesting a ten-a.m. meeting on Saturday with Henry Lipman at the plant. Tom called him back and agreed to the meeting and said that he would expand it because he wanted Peter Whitehorse and several other people to be aware of what was going on. Then he called Abbey back, who suggested that he get in touch with Stamford and Phoebe and others from the Forever Blue Public Trust Corporation to invite them to come to the meeting on Saturday.

Abbey immediately went down to the dock to find Stamford's yacht. He called him out to bring him up to date and asked whether or not he could join them on Saturday. Stamford said that he would be there and said that he would also call Tom and let him know.

Abbey agreed with Tom that as many people should be in on the hiring of the top people in the corporation. He also recommended that they invite George Rolston to participate in some manner on their board. It wasn't necessary that people like George, who had other responsibilities, needed to be full time, but they wanted them available for technical advice at a moment's notice. Abbey said that he would suggest it at the meeting on Saturday.

CHAPTER SEVEN

Priscilla Carlton, the new corporation's president, and Nancy, their secretary, were listening to Tom's conversations on the phone. When he put his smartphone down, his mother suggested a catered lunch at the plant for the meeting. Then they could continue their meeting after they were done eating that afternoon.

Tom said, "That's a great idea; can you arrange it on this short notice?"

Nancy said, "I'll get on the phone and call my friend Gladys—her brother worked for Cliffside Caterers. They often work for the police, so I'll mention Abbey's name."

Tom said, "Set it up for one-thirty."

His mother said, "That's good—we'll set it up for a dozen people."

Smiling, she looked over at Nancy and added, "We'll make sure that George Ralston is invited. The caterer can use the office across from Peter's for their setup and support space."

Nancy left. Tom said, smiling, "She'd be pleased if I set George's place right next to hers."

Tom then called Peter at the plant office and asked, "Could you provide a dozen renderings of the new plant on Saturday morning?"

Peter replied, "I'll make sure they're ready, don't worry."

Saturday morning came, and it was clear and sunny. Abbey, Phoebe, and Lieutenant Riley arrived at the plant at nine a.m. with four other officers. Only people on Abbey's list, or people on the cleaning companies' lists, would be allowed through the gate.

Abbey whispered to Tom, "I have not heard from the chief of police about the surveillance drone."

During the gathering, Tom took Mr. Kowalski and Henry Lipman aside. They followed Peter Whitehorse, George, and Lieutenant Riley on a tour of the nearly completely cleaned factory.

Tom walked up to a bench and opened Peter's drawings of their plans. Peter explained how much would be fitted into the production of the electric power. He also mentioned the advantage of the sand pit to assist and overcome the twenty-four-hour cycle of wind and solar power.

He said, "The sand battery uses sand as a medium to store energy as heat. The core purpose is to serve as a high-capacity reservoir for surplus wind and solar energy. The unit consists of an insulated silo filled with sand. It is outfitted with heat transfer pipes and tech to convert electricity from, or to, heat. Then, when it is needed, heat can be extracted from the system to provide electricity through the turbine and the generator. The silo battery will forward a hundred kilowatts of heating power and eight megawatts of capacity. The sand inside the battery can be heated to more than a thousand degrees Celsius and is typically limited by the heat resistance of the materials used in construction of a storage facility. Its storage system can be built underground, ideal for regions where real estate is highly valued."

They continued with the tour up to the second floor and ended in the entry wing where the four vocational training classrooms would be. Then they returned downstairs to Peter's office and gathered around the table that was prepared for the meeting.

Tom nudged his mother and nodded knowingly toward Nancy, who was being helped into her chair by George Ralston. His mother gave him a knowing smile and whispered, "It looks like we may have a new member of the family soon."

Tom kind-of muttered to himself, "Well, he's already on the farm."

Mrs. Carlton, as president of the corporation, tapped on the table to call the meeting to order. Her new gavel was a present from Phoebe.

There was some quiet applause.

She said, "I'll turn over the floor to our vice president and managing director, Tom Carlton, Jr."

"Thank you very much, Madam President."

Tom called on each person so everyone knew who everyone was.

Tom then said, "You are now members of the Carlton Enterprises Board of Directors, and you'll be receiving a check in the mail each month from our treasurer, Abbey, for two-hundred-and-fifty dollars."

Then he said, "As our first order of business, I'd like Stan Kowalski to give us Henry Lipman's background. Then we'll vote for his approval as our new production manager."

Stanley stood up and went on to explain, "Henry was a seasoned member of Northwest Residents' Associate, as well as having served twenty years with Cliffside Power and Light as a foreman. He has an education and practical experience in handling members of the line in any home electrical distribution."

He then asked Henry to stand up and say a few words. Henry got up, a little shy and embarrassed, and thanked the group.

He said, "I think I can do a good job getting this new green power electric company online and on the grid, to serve the people of Cliffside."

Tom then asked the group to raise their hands.

"Who is in favor of Henry taking the new position?"

All the people in the room raised their hands. Mrs. Carlton hit the gavel and said, "Action approved, congratulations."

Henry responded, "Thank you very much. I promise I'll do the best job I can do."

They all applauded.

Stanley did not sit down. When things quieted, he said, "We will be producing heat and power at a lower price than what the public is paying now. The literature I've been reading these last couple of months—" he slid a pamphlet onto the table—"predicts our green home prices will be thirty percent lower in electric and heat prices than the fossil fuel people are charging for electric distribution now."

Abbey picked up the pamphlet, and after looking at it quickly, he asked, "Stanley, what proof do they have to make such a prediction?"

Stanley responded, "It boils down to how much per MW hour cost to provide local service. A plant in Minnesota, for instance, produces service for $16.64 per MW hour because they have fossil fuels close by and don't have to pay transportation costs. While a plant in Illinois, which has to pay for fossil fuel transportation costs, has a MW price of $20.97. When the local distributor needs power, it often has to pay a higher price, to protect its profit. Therefore, they keep the prices high for the public.

"We, of course, won't need fossil fuels because we have the solar array up on the farm and can keep our prices probably around $12.50 in the northwest of Cliffside. Our competitors are not going to like that."

He looked over to George Rolston and asked, "What is your estimate MW per hour?"

George smiled and said, "When we're fully producing, we probably will be at about $10.50 a MW hour—maybe two dollars or less."

They all applauded.

Abbey asked, "How come?"

George said seriously, "No fossil fuel transportation cost. We are transporting your power underground from the array to your plant at no cost for wholesale solar power."

Stanley smiled and said, "Now is the time to play your cards close to your chest because the fossil fuel people will probably try to keep all this from happening."

Tom said, "I recommend that we make the Northwest Residents' Association and Stanley Kowalski our sales agents. All those in favor say aye?"

They all said aye.

Henry Lipman added, "When the fossil fuel people start feeling the pressure, expect a clash nationally, throughout the state, and locally."

Phoebe added, "He's right; they won't take it laying down. We'll need to be at our best to get public support. They'll probably try everything to convince the public that green power people are the bad guys. But for now, let's eat."

The rest of the day went smoothly. Tom selected a spot on the second floor on the west end for his office and suggested Henry set up alongside him. They also talked about what kind of equipment they would need.

Tom, Henry, George, Stanley, and Abbey went out to the storage trailer to see what was left from the previous owners and what could possibly be cleaned and reused. They found desks, chairs, benches, some old typewriters, and calculators—which they would replace with computers and large monitors like in Peter's office.

There were some tools, turbines, generators, and old electrical equipment, which Henry recognized but said needed to be replaced with some minor equipment.

Tom said to Henry, "Make a list of people and equipment that we will need for three shifts, and we'll start advertising that we're hiring on Monday."

Tom then asked Stanley, "Do you know more men or women who are looking for work around this area?"

Stanley said, "I'll check over the weekend and tell them to show up on Monday if they're interested."

Henry said, "I'll also investigate my old correspondence and call a few friends who were working for him on the power lines. We may pick up a couple of well-trained people."

On a clear Tuesday morning, Tom was trying to adjust to the newness of his office on the second floor in the west end of the building when he got a call from the officer at the gate. Patrick McCormick wanted to see him.

Tom answered, "Send him in; I'll meet him downstairs." Then he called Henry Lipman to join him for their first visitor, Patrick McCormack.

Tom greeted him and introduced Henry as the plant's manager.

"How can we help you?"

Mr. Pat McCormick, a tall red-headed Irish type, responded, "I knew your dad well. We were in the Lions Club together. I've seen you grow up—when I came to your farm for vegetables. I thought we should meet. I'm curious about your new enterprise."

Tom answered, "Thank you, I'm happy to meet you."

Patrick continued, "I've heard this is your new solar power electric plant, and I understand green power will be a lot less expensive than the diesel my tractors are using."

Henry interjected, "Your information is correct. Fossil fuels of all kinds will soon be swapped out with electric power—like we're seeing in some recent car models. Then, after a short time, hydrogen fuel will take over the needs of heavy equipment."

Tom added, "Henry is correct—that's the future of the heavy equipment industry. I assume you're the owner of McCormick trucking."

Patrick said, "Yes I am. How soon will all of this start to happen?"

Henry answered, "No one knows for sure, but we are all preparing.

The climate is changing, and the atmosphere needs to be cleared of excess carbonization."

Tom added, "The quicker the better."

Patrick then said, "Businessmen like me have long-term contracts with those fossil fuel providers."

Henry added, "You'd be best advised to shorten those contracts. Internationally, they're testing electric trucks now. Other heavy equipment is being tested with hydrogen fuels instead of electricity. Stay on top of it, and do your research. You're not alone—you might even be considered as a testing site for a hydrogen developer."

Tom added, "If we can be of service, please call. We're trying to prepare Cliffside for the future ourselves."

Patrick replied, "Thank you, both. I'll make my people aware of what you advise."

Patrick then offered his hand, and Tom shook it and said goodbye.

"Tom," Henry added, "one of our classrooms should teach about hydrogen."

Tom answered, "I agree, but we'll have to find a way to involve our unique sand pit and its ability to maintain heated energy for electrical storage. You'd better get out your books and do some research on these pamphlets that are being distributed. We'll have to stay on top of it ourselves."

Henry answered, "I've already found an interesting potential source. I'm going to recommend it to our attorney Raphael Sullivan. He also should investigate the new Federal Inflation Reduction Act, section fifteen: the on-hydrogen production tax credits, which offered up to three dollars per kilogram of hydrogen produced with low or no carbon emissions.

"It's a big incentive for the new ongoing projects. If we can participate with our hydroelectric brook readily available to us, it can ensure a profit and return our investment in Agrivoltaic sooner."

Tom smiled. "We're a couple of daring entrepreneurs putting all that on our investments. I'll call Abbey and have him talk to Raphael and maybe the Forever Blue Public Trust—they may back our efforts."

Henry then added, "From what I've read in the Canary Media, the Midwest Renewable Energy Tracking System (RETS) is pushing Biden to require hydrogen projects to be eligible to earn tax credits for their low carbon production. I'm sure our attorney Raphael can decipher the law for us.

"I've also read that corporate clean energy buyers such as Google and Microsoft, are working on hourly clean energy systems. RETS has a hundred-and-twenty million hourly renewable energy certificates in its international system. Grid operators who serve the United States have shown interest in hourly time stamped certificates from that organization."

Tom said seriously, "Let's put all of that aside for the moment; we have too much to do. We have to interview people for three shifts: seven a.m. to three p.m., three to eleven p.m., and eleven p.m. to seven a.m. the next morning. You said the other day, Henry, that you need three people on each shift with electrical skills. Plus, two people in the office, one custodian on each shift—and you, Abbey, and me. We'll have to have our attorney study that hydrogen information and the government offers by himself for a while."

Henry answered, "You're right, Tom. George Ralston hasn't got the array up to completion yet either, so we have the time to get prepared before he sends us megawatts of power. I'll put two men on cleaning and moving the furniture and machines from the trailer to the building—at least, the ones we want to use—and we'll just dispose of the rest or see what we can get for them."

Abbey, who had walked up in the middle of their conversation and was quietly listening, added, "I'll stay on top of security and see the money people at Forever Blue. Phoebe will probably come with me every day. We'll be setting up the vocational classrooms and an espionage area. We'll advise board members from the Board of Education through her friend Stanley. It would help a lot if we had public recognition for our vocational effort."

They all laughed.

Abbey then said, "Don't forget, Peter Whitehorse is bringing in his metal people next week to construct and insulate the steel sand pit and installation for the 5,330 cubic feet of sand."

Henry said, "Will I ever get home?"

Tom's phone began ringing. He excused himself and walked away a few paces while listening. A few minutes later, he returned, saying, "My mother needs me at the farm store. You guys carry on, and I'll get back to you as soon as possible."

He then left, got into his car, and drove out the parking lot.

Tom arrived at the farm, walked into the store, and greeted Bill O'Donnell, his late father's closest friend.

He said, "I haven't seen you since I went way to school."

They shook hands, and Bill answered, "You're looking fine, Tom, and from what I hear, you have hit the ground running since you graduated as an electrical engineer. I'm now the chairman of the Northern Farmers' Association, and we need to know what you're up to. It has been our intelligence drone you've seen overhead and have had the police threaten our arrest for invasion of your privacy."

Tom asked, "What do you want to know?"

Bill answered, "What is Agrivoltaic?"

Tom took him aside, saying, "You haven't been doing your homework. Agrivoltaic is a recent development by a lab at Arizona University, combining agriculture and solar panel power. They have found mutually beneficial connections—it's symbiotic."

Bill asked, "How does it work?"

Tom answered, "Come with me, and I'll show you."

He called George Ralston on his smartphone and alerted him to his coming with a guest to show him the array of Agrivoltaic.

George answered, "Come on in—I'll give him a tour and give him my favorite introductory speech."

Bill said to Tom, "I have a couple of men in the car, can they come?"

Tom answered, "Sorry, but this is a top-secret development. Bill, you can come, but you'll have to keep the details to yourself for about six months. We'll be going public then. "

Bill said, "Okay, I'll follow your instructions and tell them I'll be right back."

Tom led him through a channel in the fence and then walked him across the field to meet George. Tom introduced Bill to George as his longtime, late father's friend, and the chairman of the Northwest Farmers' Association. Then Tom introduced George as the electrical engineer and construction chief for the Morning Sun solar panel system. They shook hands, and George said, "Let's take a walk into my forest of steel."

Bill followed, saying, "This is acres of upright steel posts."

George smiled. "You'll notice that they're fourteen feet high in straight rows. We'll mount solar panels on each of them, all forty-five acres. They will produce six megawatts of DC power, which we'll convert to AC green power, and Tom will sell it to the local grid at roughly a third less than the local grid pays for fossil fuel power now. This will give Tom a nice winter income, as well as build a nice future investment.

"You'll also notice the rows are thirty feet apart and the workers are plowing the earth between the steel rows for the planting of Tom's crops. They will grow faster and taste better having been grown under the solar panel's protection, which rotates the sun's rays to get the best growing angles each day."

Bill turned to Tom and said, "Your dad would be proud of this."

Tom answered, "I'd like to think so."

Bill said, "It'll take a hell of a sales job to outsell invested fossil fuel people and the 'not in my backyard people' as well. People, even now, are resisting green power all over the county. I'll do all I can to convince our associated farmers that solar panels and farm crops can successfully benefit each other in this symbiotic relationship and make sure they realize it's all carbon free."

Tom added, "Let me give you a part of a quote from the researchers in Arizona: *'Growing agricultural crops under the shade of solar panels uses less water and is much more efficient at shading plants from the worst of the midday sun. On one Arizona site, cherry tomatoes doubled in size and required less water when grown in the shade of solar panels. This implies an important increase in a farmer's income.'*"

Tom continued, "As I've said many times, it's taking advantage of a new method, creating a surprisingly robust source of a farmer's income."

Bill said, "Thank you," to George and then followed Tom back to the farm store.

They noticed the car was empty as they approached the farm store and looked around to see what had happened to the two passengers. Tom smiled and saw them sitting at the farmstand's welcoming table.

Tom said to Bill, "My mother has had that table for twenty years, greeting whoever comes to the farm store or to buy their vegetables. Bill's passengers probably got curious about the famous Carlton vegetable farm store. It's been feeding Cliffside families for over a hundred years."

Tom and Bill came up.

Tom said, "I see you've met the boss, my mother, Priscilla Carlton."

They both stood up.

Priscilla said, "They were curious about the market but more apprehensive about our new solar panel array. I know just how they felt; at first, I was quite apprehensive myself. But the villains in our time are not those people erecting solar panels on their farms—as we have—it's the 'not in my backyard people' and the changing climate. I'm not ready to fight carbonization."

Priscilla continued listing the problems we are facing and have faced.

"A worldwide pandemic, the extremes of flood, drought, the never-ending downpours, as well as wind and dust storms. They have all been frightened by the changes, and I'm not just saying this country, I'm talking worldwide. We're facing an epidemic of the shooting of school children and most of all, attacks on the capital we love.

"We need to reach down and draw up the strength of our forebears to grasp the tools of the new world. This includes smartphones, the internet, and artificial intelligence. Reach down and create new messages and new rules of survival. We need to fall back on our foundations of family, community, strength, and the political constitution that we have had over the last two-hundred years."

Tom laughed and said, "I say with pride, Mom, you really need to run for Congress."

"Oh no," she replied. "Like you all, I'm just a product of the land that we live on."

Bill turned to his passengers and said, "Now fellas, let's go back and tell them all it's a new source of power and income that has a symbiotic relationship between solar power and agriculture. I have just seen the Agrivoltaic setup that will give us the power of the sun to help us meet the power of new climate changes, not from carbon-infested fossil fuels but by using green power from the sun and, perhaps soon, hydrogen."

CHAPTER EIGHT

Several productive weeks had passed since Bill O'Donnell's visit to the farm. The new employees grew comfortable in their new jobs. Henry Lipman ordered and put in place new electrical equipment, sand pit was filled with sand, and the array on the farm was complete and in the final testing stages before it sent power to the factory.

The management on the grid that served Cliffside agreed to accept all power from its northwest substation. They agreed to accept the new green power and to try out the heat storage sand battery for its excess DC and AC power.

Phoebe Sinclair had, with the help of a member of the City Council, approached the Board of Education about the Green Power Factory offering vocational classes to adults.

They all awaited the call from George Ralston that Agrivoltaic was ready to send solar power and grow vegetables for the market.

Mr. Joe Hendricks was owner of a half-dozen gas stations at and around Cliffside and Bayside. As usual, his three buddies had a game of cards on

Wednesday evening in his Coolidge Street recreation home in Cliffside's posh east side.

During the evening, Joe brought up the growth of green power and solar panel installations in the county: on the farmlands and on people's home rooftops. He felt it was becoming a threat to his gas stations, because he had a long-term contract with his fossil fuel suppliers.

Phillip Stevenson said, "I'm also feeling the change in my food stores. Food suppliers are being pressured to buy their vegetables from Agrivoltaic farmers because of their improved symbiotic growth. They're not buying from the usual local wholesalers of fruits and vegetables but from independent farmers in their private vehicles."

Patrick McCormick said, "Fruit and vegetables that are trucked in from the West have also slowed down."

Councilman John Coolidge added, "Things are changing, and people worry about the carbonization of the air. We must be prepared to make the changes with them. Fossil fuels are on their way out because they create carbon. The future will probably not be gasoline fuel but electric charging stations and electric motors in vehicles and trucks."

Patrick then added, "I've been doing my homework, and my big trucks and heavy equipment won't have diesel or electric power anymore—it'll be electric or hydrogen carbon that'll be moving the vehicles."

Councilman John Coolidge then replied, "It hasn't affected the city budget just yet, but it soon will. The councilman who has the electric grid in his district tells us it won't be long. The Cliffside electric grid distributor is looking forward to paying less for green power than they're paying the New England grid fossil fuel supplier. They'll be happy to pass the lower costs onto the homeowner."

Patrick added, "The Carlton Farm Enterprise Corporation is making a major investment in solar power, which will be at a very competitive price. Maybe we should look at our long-time fossil fuel contracts."

Phillip then said, "Maybe we should ask Tom Carlton to slow it down?"

Patrick said in response, "The old Tom is dead. The son, Tom, is a graduate electrical engineer, and he's got major financial backing. He's putting in forty-five acres of Agrivoltaic panels that will produce enough power to supply all the houses in the north side of Cliffside that are presently served by the grid's substation. They'll buy cheaper power from him."

Joe agreed with Phillip. "Then we should ask Tom, Jr. to slow down."

Councilman John and Patrick answered together, "The Carlton Family Farm has been an important fruit and vegetable source for a good many generations. Old Tom carried us all during the depression in the thirties. We wish them well, and we don't support anybody trying to give them a hard time."

Phillip and Joe denied doing anything to hurt them, and Philip quickly changed the subject. But he gave Joe a reassuring look.

Later, Phillip asked Joe, "What's the symbiotic thing they're using to grow bigger vegetables?"

Joe answered, "I don't know much about it either."

Phillips said, "I'll look it up and get back to you."

They changed the subject and got on with their game and talked about the upcoming football game between Boston and Philadelphia.

Later that evening, Joe sat thinking about the loss of his gas and repair businesses that were made possible through his fossil fuel connections. He said to himself, "I'm not giving up. I'll get together with Phillip, and we'll develop some kind of plan. I'm sure McCormick will move to help the Carltons, and they're developing Agrivoltaic that's symbiotic—whatever that is."

Meanwhile, back at the Green Power Plant, Abbey was talking to Tom and Henry.

"I've approached the police chief. I asked that his detectives assist in the espionage defense against the fossil fuel people. The chief agreed and ordered all the detectives to watch and listen on their microphones that are placed around town for any conversations or actions related to green power, solar panel electricity, or Agrivoltaic vegetables."

Several days went by before word came back that the grocer, Phillip, and the gas station owner, Joe, had met to discuss the Carlton's new improvements, especially in regards to the old coal plant.

Stan Kowalski of the Residents' Association was developing a green power sales office for the sale of green power solar panels and Agrivoltaic installations by the country's farmers, as well as home roof solar panel installations. He asked a new resident employee if he could get a tour of the Green Power Plant. Stan agreed. One of the policemen on duty heard about it, reported it to Abbey, and he took it to Henry.

They arranged to have a Police Department electrician act as his tour guide and that he would wear a bodycam. They gave Stanley an okay to take him on a tour with the policeman. No one was to know about the arrangements except Henry, Abbey, and the policeman.

Stanley gave the man and the policeman a grand tour of the plant from top to bottom and a free lunch. He even agreed to submit his camera at the front gate.

Following the tour, a private investigation was carried out by the Police Department and provided to Tom, Abbey and Henry the next day. The final conclusions would be left to a special meeting that Tom called after the tour at seven p.m. on Tuesday evening.

After the plant tour, the guest asked to see the solar panel array at the farm. Stan asked Abbey if it was okay, and Abbey said, "Sure, go ahead, but make sure you take the bodycam with you."

Abbey decided that after they left for the farm, he would seek out

Peter Whitestone. He asked Peter to send one of his men over to the reservation lab and to tell George to use his bodycam to record their visit. The rest of the crew would want to see the footage the next night at the staff meeting.

When Stanley, the guest visitor, and the policeman arrived at the farm parking lot, they discovered George Ralston had moved the fence on the west side of the farm back two car links and that he had installed four electric automotive charging stations. They were painted a bright green. On the back of the fence post, a sign invited electric auto owners to use their Visa cards and to drive safely.

Stanley laughed and told the policeman to get a good photo. Priscilla, who had Phoebe visiting, described the tour and invited her to join the meeting. She called George on her smartphone, and he suggested they drive down the road to the main entrance so he could meet them by the panel office gate.

Abbey called George and alerted him that they were coming.

Stanley drove down the main road to the solar array entrance, and a guard opened the gate in the fence. Stanley drove up into the parking lot. They all got out, and George introduced himself then proceeded into his usual spectator speech.

"Agrivoltaic has a beneficial symbiotic relationship with vegetables. There's a new development by the University of Arizona explaining how the soil is protected from the heat of the sun—how the panels trap the heat to preserve groundwater for the plants.

"It is quite simple. Protons under the panels reduce the surrounding temperature, lessening the heat island effect of the panels as the water transpires through the plants and then are cooled during evaporation.

"The farmer benefits from better and more abundant crops and a nice new income from the electricity produced, and even more so if he drives electric farm vehicles and personal cars."

They toured several parts of the installation and finally returned to the gate and the visitor's car. They all shook hands with George and made their way back to the plant. The visitor thanked them for their time, the tour, and lunch, then left for his home.

The sound and views from the concealed bodycam were delivered to Tom's office then given to Peter's guy, who carried it over to Forever Blue's lab on the Reservation and to the manager of Morning Star for a sound conversion. Then the videotape would be returned to the Green Power Network Plant by a messenger at about four p.m. the next afternoon.

Henry, who was in Tom's office at the time, asked about the messenger. Tom said, "Our sponsoring corporation, the Forever Blue Public Trust, owns and operates a bodycam security laboratory on the Abenaki Reservation. The messenger works for them."

Later, Tom's senior members' meeting happened at seven p.m. The meeting included Nancy, his sister; the corporation secretary, Henry; the general manager; Peter Whitestone, president of Wikiup architects; Abbey and Phoebe from Forever Blue Corporation; George from the Morning Sun Corporation; Stanley from the Northwest Residents' Association, and the new manager of the Green Power Network's sales office. The Cliffside police lieutenant, as head of security, joined them as well.

When they quieted down, Tom called their attention to the four-by-six mounted monitor at the end of the room. Peter slipped in the cassette on the computer and the bodycam video began on the monitor and it ran for forty-five minutes. The group was impressed as it showed all of the subjects, especially the visiting guests.

Tom said when it ended, "How about that. Thank you for our cooperative new video department presentation."

They were all impressed by the quality and the depth of the sound. Most of them had made notes as it had played, and they especially complimented the policeman who used the bodycam with real dexterity.

Abbey said, "I'll convey your compliments to our chief of police."

One at a time, Tom went around the room, listening and discussing the areas the visitor was interested in. Abbey and the police lieutenant took notes.

Tom asked, "Where are we the most vulnerable?

Abbey responded first. "The floor with the turbines, generators, and grid connections. Second, the water conversion tubes from the brook to the sand pit. Third, the east wall, outside of the insulation, and the steel sand pit. And lastly, the farm's solar panel array control building and the transformer connections themselves."

Tom asked, "Do we all agree?"

They all indicated their agreement by saying aye.

Tom turned to Abbey and the police lieutenant and said, "That will be your joint challenge and assignment: Find a way to protect us."

Tom focused on Abbey. "It is abundantly clear why Forever Blue Public Trust chose you as their corporate treasurer. You're very meticulous in your analyses."

Abbey said, "Thank you. But first let me tell you our intelligence service has learned our guest on the tour is really Steven Robotham, the field representative for New England Oil and a gas company of Boston. He is the area's fossil fuel supplier, including for six of Joe Hendrix's gas stations and his other automotive supply businesses. Apparently, they're curious about Morning Sun's solar panel activities in the northern New England area.

"Stanley, I would not recommend that you give Steven a job in sales. We know who he really is. Also, since we'll be carrying the Morning Sun's green power line in our store, we should tell George's boss what we have learned."

Tom spoke up, "Good idea."

George interjected, "Okay with me; I'll call him tonight."

Abbey then said, "Ask Mr. Donovan to please keep this information to himself."

Henry then stood and spoke up.

"Before we go any further, we need to make our electrical mission clear to the New England and the National Grid branch; our basic mission should be published through our sales office. The basic mission of the Green Power Network of the Carlton Enterprise is to protect vulnerable citizens in Cliffside by making the following our priority:

"In the event of a civic emergency such as wind, water, fire, or any other natural disasters that cause a power cut off, our green power will be available to the public without charge.

"We will also provide power under the same circumstances to the following: police, firefighters, EMTs, hospitals, Meals on Wheels, the Salvation Army, the Red Cross, senior housing, and their transportation.

"We recommend that all of our home service customers and salespeople to be equipped with generators and sand energy storage for up to seven days in case of such an emergency.

"If there are objections from the board, let's hear them now."

The room was deadly quiet. It was obvious they sensed their general manager was talking from years of home electrical distribution service experience. They did not want any misunderstandings.

George spoke up and said, "Morning Sun does not currently provide or send emergency power with solar installations."

Tom interjected, "Ask Douglas if he'll develop it for our Green Network for home sales. We'll pay Morning Sun for its development, along with the forty-five-acre array.

George answered, "I'm okay with it, and I'll do my best to convince Douglas. It's a long step forward for private home power—seven days of power with a house generator. Wow! We'll be the ones leading the country."

Tom looked over at Nancy. "Have you gotten all that?"

She answered and said, "You bet, every bloody word."

Phoebe spoke up, "I've also recorded this whole meeting. I'll have it typed and sent to each of you. The Green Power Network will be distributed to seniors and their future home power, as well as each member of Forever Blue."

Tom cautioned, "All of you—keep this to yourselves. It's a major challenge for any corporation, and we are in a big and very tough game."

CHAPTER NINE

Ezra Woods, a longstanding City Council member and a farmer in his late seventies, had been a gas station customer of Joe Hendrick's for many years. He was cornered one morning by Hendrix, who asked, "When you go back to the City Council, help me keep this Green Power Network from putting me out of business. Convince all the people, especially the farmers, from moving to electric-powered farm machinery and automobiles."

Joe's opinion was that the Green Power Network would resist a trade. Ezra would introduce a bill to the City Council to disapprove the request by the Green Power Network. In order for them to get the upcoming nonprofit, public service corporation permit, a committee would have to study its purposes.

Joe knew that the network would not be able to supply power to the grid while the study went over their solar power sales to other potential customers. It could take a considerable length of time.

Ezra thought it a good idea and agreed to introduce the bill at their next meeting. He was assured by some fellow councilmen that they would support it.

Joe then called Phillip and told him of Ezra's bill. He also asked him to put some pressure on the City Council.

Phillip said he'd call Patrick McCormick and ask for his support for the study committee bill as well. The thought came to him that Patrick McCormick would probably not support the bill, because he was a friend of the Carltons. Phillip agreed that it was a clever move on Joe Hendrick's part, using the aged and naive Ezra Woods as his City Council connection. He felt that John Coolidge would be against it, though, along with Councilman John Stevenson and Carmen Spinoza. But they would need four more additional council votes to kill Ezra's bill.

Bill O'Donnell, who was the chairman of the Farmers' Association, would probably throw all his power behind the Green Power Network request for an approved permit.

Many people who were in favor of fossil fuels and its carbonization of the atmosphere would support the bill to study the Carlton Enterprise. However, members of the Northwest Residents' Association and friends of Stanley Kowalski would contact their people and see if they could be assured of their votes for Green Power.

There were at least two additional votes on the council from the Dairymen's Association. Tom would have to talk to Bill O'Donnell and see if they could be persuaded to vote with the Farmers' Association. The leading members of the Dairymen's Association were Calvin Russet, who owned a hundred cows and was influential in their association, and Archie Moore, who had seventy-five cows.

Tom got together with Henry at the plant to review what they knew of the dairy business around the county, trying to figure out how to approach them. Neither of the men had solar power in their dairy operation.

Bill mentioned, "They've been having financial difficulties because of the price of milk going up and down on a weekly basis. In the last couple of years, it's been a get-bigger or get-out mentality in the dairy business in New England. The dairymen who have a thousand cows or more seem to be secure, but the family-sized dairy farm, with just a dozen cows or up to fifty or sixty—between the costs of electricity and milking machines and the selling price going up and down, it makes it a difficult market for traditional family dairymen."

Henry said, "As far as I know, the average number of cows per farm is between a hundred-thirty-eight and two-hundred-twenty-eight through-out the county. Any dairyman who has less than that is facing hard times and will be looking for outside help. The hours of a dairy worker are long, and it's a physically demanding job. The pay is low, and they have no labor laws to protect them. When the price goes down, they increase production to be more efficient, which worsens the situation. Even with federal pricing assistance, most of the dairy farmers under that level are steadily selling out to the real estate developers."

Bill finally said, "I believe Agrivoltaic installations on these large dairy farms would give them the additional electric assistance from solar power installations, especially with the size of dairy farm roofs and the milking machines needing power twice a day. It will also make up for seasonal dairy income shortages. The extra electric power they provide from their barns, homes, and converted vehicles will be a boost to their winter income. It has been a big help to most of the regular farmers in my association."

Henry said, "If Peter and his architects could create some renderings and income figures, we could show them to these dairy farmers during the City Council meeting. Perhaps we can get their support and have them join in with the regular farmers. I'm going to call Peter right now to explain it to him and ask if they can help."

Tom said, "Good idea—tell them I also support it."

Henry came back off the phone and said, "Peter will get to work on it right away. He thinks it's a good idea, and they'll help as much as they can. It will take them at least a couple of days to produce anything that they could use at the City Council. However, he said the dairymen are pessimistic about their future. Most of them can't afford to get bigger or get out. The public could certainly help by being willing to pay higher milk prices, but that's not likely to happen right away."

Henry said, "Not necessarily. I think what they need is better management, like solar electric power in dairy operations—for milking and so forth. We'll place panels on barn roofs, and the access to electric money in their pockets from future grid sales could quickly brighten their future. I think we're just the ones who can do it for them."

"That reminds me," Tom said. He picked up the phone and called George Ralston at the solar installation. When George picked up the phone, Tom said, "Please run an underground cable from the stations at the farm store to the wall of our big barn. We're going to get all our diesel vehicles electrified by McCormick as quickly as we can."

George said, "Will do."

Tom hung up, turned, and said with a twinkle in his eye, "How would we like to have an electric-powered robot cultivating the weeds away from the vegetables?"

Henry looked at Tom with surprise. He asked, "Where did you hear about that?"

Tom replied, "I read an article about an engineer working in California for an open-field farmer who was experimenting with one."

Henry said, "That's going to put a lot of crop pickers out of work, as well as regular farm help."

Tom answered, "Not out of work—there will just be retraining on new equipment. But it isn't going to be tomorrow. Pat McCormick better start reading up on AI electric changing over to robots for heavy equipment."

Tom continued, "Bill O'Donnell, president of the Farmers' Association, has spoken with its farmers who have memberships and has invited the county's dairymen to join with the farmers. Also, he's going to ask the dairyman to change their name to the United Dairymen. I'm also going to ask my mother to add a wing to the farm store for dairy product sales.

"Morning Sun has agreed to sell solar panels for the roofs of the dairymen's barns to provide them with electric power for their dairy machinery, as well as their financial needs. Our Green Power Network will appeal on their behalf to the state and the federal government for sources of income. Their power processing will go through our network sales office, and their dairy product sales will be tracked through the farm store. It will be nice to have us together and politically stronger in dealing with the public, the fossil fuel opposition in the City Council, and elsewhere."

Henry said, "Sounds good, Tom. I'm sure our Residents' Association will agree."

Tom stepped behind his desk and reached for the phone. "I'll call George to discuss investigating the idea of a robot clearing weeds from growing vegetables under the solar panels on each acre."

Henry shook his head, saying to himself, "God, give us the strength to live with Tom's intellectual curiosity and entrepreneurial drive."

Tom smiled and said, "My generation has to match the great generation that preceded us."

George answered the phone. Tom asked, "Are you familiar with the effort in California to create robots that can pick weeds from between the growing vegetables?"

George answered, "Yes, I read about it in an article in a science magazine, why?"

Tom asked, "Are they advanced enough for us to buy a couple to use experimentally on our forty-five acres?"

George answered, "I don't know, but I'll ask my boss to inquire without using our green network name."

Tom answered, "I'd appreciate that. Keep me informed."

George responded, "I'll call him right away, and I'll get back to you when we get some kind of real response."

Then Tom called Raphael Sullivan, their lawyer, and told him, "there are concerns with the City Council and getting the dairymen to join the network, and there will be a need for state and federal solar panel installation assistance."

Raphael said, "I'll get on it right away. I have read the state and federal regulations and how they might affect our new partners at City Hall. I've begun to understand that Ezra Woods suggested a bill to the City Council. He also reported having heard some whispering campaign about an unknown ingredient that may be dangerous in the vegetables that we're growing."

"The word they're concerned about," Tom advised, "seems to be "symbiotic," which the grocer is pushing around. We'll need to explain to the public the advantage of solar panels shading the vegetables and providing more water. I found the process to be a positive symbiotic relationship, and it's not a kind of chemical ingredient that could harm the public."

Just as Tom was hanging up with Raphael, Pat McCormick walked into his office. Tom greeted him with a friendly handshake and invited him to sit down and asked, "What brings you back to this office?"

Pat replied, "I have been doing a lot of reading on electric vehicles, especially trucks. And I would like to ask a favor. Will you allow me to electrify your farm trucks and other vehicles? I'll do it for free to train my six mechanics in handling the conversion from diesel and gas, to electricity. I would like to have my men have the time and patience to successfully train themselves in the conversion, so that I can advertise the new service to my customers who have heavy electrical equipment and trucks."

Tom called George up at the farm and asked if it would inconvenience him. George responded, "We can work out the schedule so he could work at convenient times. I'm sure we have enough power at the recharge stations near the barn to handle the equipment after it has been converted to electricity."

Tom called his mother on the phone to explain it to her and asked how she felt about it. Mrs. Carlton replied, "I am in favor of the conversion. I thought it was the trending thing anyway, and it would give us a positive jump on the market. Also, Tom, I'm having a family supper this Friday night. I'm inviting Phoebe and Abbey, so put it on your calendar."

Tom said,"Okay."

Then he told Pat her reaction, and they made a deal. Tom asked, "How soon do you want to start?"

Pat replied, "Right away."

"Okay, Pat," Tom responded. "Our plowing is done, and the fencing is about finished, so let's do two tractors first."

Pat said, "I'll send my tractor over to pick up your two vehicles tomorrow. I don't know how long it'll take. They'll be doing it in their spare time between other jobs, but I'll bring them back as soon as I can."

Tom replied, "No problem. If you need electric type help, call us and we'll help you with the engineering."

Tom then asked Pat, "Do you provide parking for the school district buses?"

Pat said, "Yes, why do you ask?"

Tom responded, "Let me read you a quote by Paula Morrows from WBUR, in Beverly, Massachusetts, in relation to school buses."

Tom opened a pamphlet on his desk and began to read: "*Summertime in New England is when people demand the most electricity from the grid because of air conditioning use. Utilities turn to 'picker plant' to supply extra power. They're often chosen at older, more polluted facilities that are more expensive to run. But a project in Beverly, MA, offers an alternative on demand power source. The school district uses their electric school buses as giant batteries and has power plants that send energy back to the grid.*

"The project uses file direction chargers that can charge the bus battery and allow the battery to send energy back to the grid. The front panel of an electric school bus is the charging station and shows the direction of the charge as well as the battery level.

"Last summer, the project increased from one bus to two buses, and it sold seven megawatt hours back to the grid. Electric buses are ideal backup power for plants for two reasons: First, they carry big batteries called lithium-ion batteries with three times as much capacity as an electric car. Second, the school buses usually don't have a summer job.

"Hypothetically, 9,000 of Massachusetts' school buses were electric. National grid estimates they could practically contribute 450 megawatts during peak times, equivalent to powering 375,000 homes.'"

Tom exclaimed, "How about that?"

Henry, who was listening from his desk said, "Tom, we can easily produce what Beverly buses provide now with our sand energy battery. That may be something to consider to improve our summer business."

Tom smiled and said to Pat, "Check the buses out quickly, and maybe it could be a motive for you to convert school buses as well as trucks."

Pat thanked Tom and Henry sincerely and left the office to get the job done.

Later that day, Tom, while sitting at his desk, received a call from Stamford at four-thirty p.m. Tom greeted him happily and asked, "How are you, and what brings you on the phone at this time of day?"

Stamford replied, "Got lonely for the family, Cliffside, the bay, and

I felt like putting my yacht to work. I'll be arriving at your place in less than a half hour."

Tom responded, "We'll be glad to see you; it's been some time. I'll call my mom and have her put another plate on the table for supper. She called earlier and said she was doing something special."

Stamford replied, "See you shortly."

Tom called his mom at the farm and told her about Stamford. He then asked her to set another plate for supper.

She said, "Oh, that's wonderful—the whole family will be here."

Then Tom went out to the main gate to wait for Stamford to pull up in his car. When he arrived, Tom got into the car and told him his mother and Nancy were waiting for them.

Stamford laughed. "Priscilla is always looking for any excuse to have a big meal."

He turned the car north then said, "Good, I can use some good old family small talk. It's been just business talk for weeks on end—ever since I joined my wind power to your Green Power Network. You and Henry have been busy, demanding night-time power to keep your sand-pit energy at a thousand degrees for your daytime power-grid business needs. I'm making a lot of money, and the Forever Blue Trust is pleased with how fast George is working—acres of power have come online. But I'm happy to be away from it for a while."

Tom answered, "Your night-time power boost has helped our startup time immensely. You'll have to come by tomorrow and watch Henry in action with his six-man team that produces heat to the turbine gener-ator that generates power—also heat water from the stream from the back of the building and hydroelectric power from our roaring brook."

Everyone else was already there and seated as Priscilla and Nancy welcomed Stamford. They talked about the weather with him, covering everything from the ocean weather to his latest clothing purchase—like the style of the suit he had on. Even Nancy's close friendship with George Rolston came up.

Priscilla said, "Now that the cat is out of the bag, Nancy and George have announced their engagement to be married in about a year."

Stamford smiled and said, "Oh my!"

Nancy was a little embarrassed but admitted they had been seeing each other for a while. That was the reason for the special supper with the family.

Stamford teased her a little by saying, "Well, it is just a short way across the grass to his office."

They all laughed, and Nancy blushed.

Stamford, showing some sympathy, then said, "Tom, I think this chicken is nice with olives, don't you?"

Tom picked up on it and agreed, "The food is always very good here."

Priscilla helped Nancy out too and questioned, "By the way, Stamford, what's the ocean like this season?"

Stamford answered by saying, "It was smooth as glass coming over. I'd love to have you and your team take a trip to my farm; the weather has been good and will be for a little longer."

Tom said, "We would love a day on the water; it would be a nice holiday. We've all been working hard."

Stamford said, "Before I go back, we'll set a date, I can handle eight or nine people besides myself and the crew, if that's okay."

Tom turned to Abbey and said, "Check our schedule for a Sunday in the future when our senior staff can make it. Then call Stamford and get back to him."

Abbey answered, "I'll get together with Henry and set it up."

Stamford put up his hand and said, "No more business tonight—let's enjoy our supper."

The dinner went well. Tom's mother was always a fine cook, and she had prepared a poached salmon with a wine sauce, chicken with olives, potato soup, and mashed potatoes and mixed vegetables.

Abbey commented, "I have never had such a marvelous potato soup; it was beautiful with the bacon on top. Priscilla, thank you, we're going to have to invite you to the Painted Lady for one of our dinners."

After dinner, dessert was pecan pie and coffee in the living room of the colonial home. Nancy had started a fire in the fireplace. In the quiet, Phoebe took the opportunity to sit next to Tom to watch the fire.

Phoebe asked, "Tom, I'm a conservative, seventy-five-year-old lady, but between us friends and family, I'd like to know why a man of your ability and education would put this colonial home on the line to venture into the development of green electric power for a nonprofit public corporation? Feel free if you would like to tell me—it's none of my business, but I'd surely like to know: What drives you this way?"

CHAPTER TEN

The room quieted, and all eyes and ears were on their leader, Tom Carlton. Nancy moved over next to her mother, who was obviously emotionally apprehensive over to her son's answer. Tom sat quietly, reached over, and put his hand on Phoebe's.

He said, "It has to do with the fact that my father developed pneumonia in the later part of his life and could no longer function as the beneficiary of this northwestern area.

"He was conscious of the history and our family's responsibility to the growth of Cliffside. I pledged to him that I would adopt his code. If you know French, the expression is *noblesse oblige*. My father took it to mean that the rich and the successful had to be responsible for the livelihood of their fellow citizens. His illness prevented him from continuing in this capacity in our community. On his deathbed, he asked me to continue on his behalf. I pledged to him that if I was successful in my college career, I would continue to serve our community the best way possible. My education has taught me to know that the solution is the conversion from fossil fuel to carbon green electric power."

Phoebe responded by saying, "I am the president of Forever Blue

Public trust, and we're pledged to public service. I shall do all I can with their support, to give you our best financial advice possible."

Stamford was the first to speak.

"You know that Abbey and I are also both members of Forever Blue Public Trust, and we, too, pledge our support to our president in support of you and your green power venture."

Tom answered, "My family and I are grateful for your support. I know that our financial health is secure because of your backing. Thank you.

"Henry and I got together the other day and talked about security at the plant and the controversial people being approached by one fossil fuel person in Cliffside. They have been creating a whisper campaign against our solar power growth, and they're pushing a bill in the City Council to hold up our network's zoning permit and to private home solar panel sales."

Stamford held up one hand and said, "No business worrying about it tonight. I'll help Abbey, but we'll get to it tomorrow, okay?"

Abbey answered, "Our true growth adversary is not the fossil fuel producers but a twisted informant who has just been introduced into the community. His specialty is turning legitimate information into subjects of business that might confuse or cause customers to go elsewhere. We need to protect our customers from misinformation of political design. We need to protect our basic capabilities.

"Our real competition is the influx of dark money. The people doing the work of the fossil fuel people are being paid under the table to prevent our green power distribution to private homes and businesses that can use the additional financial advantages—including our new friends in the dairy business.

"The people asking questions, that are wanting to visit and have a tour, are going to be quite ordinary, but they probably will have specific intentions. We need to know who they are and when they're going to show so that we can be prepared.

"New businesses trying new ideas need room and protection during the developmental periods. Through the influence of dark money, people might be hurt, unless our green power replaces their fossil fuel power. I'll hopefully be able to get the police involved and get back to you."

Tom asked, "what about the government helping?

Abbey waited a moment then said, "Even though the European Union has taken major steps towards controlling dark money and the direction of AI, the United States is still floundering in its attempt to control them."

Tom put his arm around Abbey and said, "I thoroughly agree to your idea of creating defensive tools for our protection and the good purposes we serve. Tell me specifically what we're going to need for staff, whenever you're ready. And don't forget to call Stamford and set up the yacht trip."

Abbey said, "I'll contact the chief of police and make sure we fit into his program, add some incentive maybe by making additional equipment available to him. I won't forget to call Stamford."

Abbey then left for the snack room. When he came back, Mrs. Carlton spoke up, "I'm going to suggest as your president that I'd like to see something added to the articles of the incorporation. Can I do that?

Abbey said, "As our president, you can propose whatever you like, then the board will vote on it."

She answered, "Good. We have agreed with the grid to reserve our power in the case of an emergency and to distribute it to emergency services like police, firefighters, hospitals, and people in real trouble, but we have not mentioned EMS volunteer services like those out in the country and at the Abenaki reservation. I believe we should have them on our list just in case of an emergency."

Abbey asked, "Why do you bring this up now?"

Priscilla answered, "At this time, there are only two full-time volunteer paramedics driving one old truck around our sparsely populated county, and they don't go into the Reservation, to my knowledge. Meanwhile in Cliffside, we have a fully paid medical emergency team that won't go out of the city due to City Council's regulation on paid taxes. If we're truly going to help the needy, let's do it now."

Phoebe said, "She's right; if the Green Power Network is going to live up to what it has claimed, we need to start helping people."

Henry overheard and added, "I can recall having helped two guys and a girl who were trying to get a seriously ill older man to the hospital twenty miles away, and they weren't able to because all the gas stations were closed. If they had a full tank of gas in the truck, that man might be alive today."

George Ralston spoke up, "Mrs. Carlton, no one here is against it, but you're talking about two hundred-thousand-dollar EMT trucks and two men to work those trucks twenty-four hours a day, in both our eastern county and our western county. The Reservation, I believe, has something they're using for that purpose in the eastern county, but the western county being so sparsely populated is why they only have one old vehicle."

Stan Kowalski spoke up, "Our Residents' Association has often thought of it, but we could never afford it."

Phoebe added, "We have good friends at the Reservation and at the local University—I'm sure they could provide some volunteer help."

Tom overheard all of this from his mother and said to Phoebe, "Contact the University and the Reservation Control Board. Tell them we'd like their assistance in setting up EMS equipment in our eastern and western county. Also make sure to ask how they feel about it."

He continued, "Abbey, check your senior list. Let's see whether we have any older people who might want to help with four-hour shifts in the east county.

"Stanley, contact your people in the association for four-hour shifts as well, asking for volunteers in the western county."

Tom continued, "Stamford, you and I are going to see Pat McCormick and buy two EMT trucks. We'll give him ten thousand each on our credit card in addition to the task of changing his vehicles to electric power."

Then Tom turned to his mom and said, "We'll give it a good college try, Mom."

She sat down next to Nancy, clearly comforted.

Tom called and asked Raphael to report back if the board voted yes to help EMS in the east and western counties.

Tom then asked Stanley, "How is it coming with the Association for our home sales solar panel office?"

Stanley sat back in his chair and said, "Henry and I have been pushing other members of the Association to devote more time and space to convert our warehouse into a sales room and office to accommodate the solar panel samples and equipment for home sales that are being provided by Morning Sun. George has helped us with the space and the electric connections that we must have. Henry is also a licensed electrician who will wire the connections when the carpenters are finished.

"The Morning Sun has agreed to deliver solar panels and other equipment in two weeks to use for demonstrations, along with instructors to help us learn how to use them. We have two men and a woman prepared to be salespeople. We should be open for business by the first of the month. We also have five families that are ready to purchase the resident homeowner's equipment. I think it would be nice to offer these fine people little bonuses as being the first five to purchase."

Tom looked at the group and asked, "When will our company take over their maintenance?"

Henry said, "I think the network should cover all maintenance costs for the first year of all sales. That way, the network will be assured that everything is working in good order and moving forward smoothly. Our company is working on the emergency sales plan, a seven-day insurance reassurance that the home will not lose power if the right equipment is in there."

Tom said, "I'll have our lawyer find out how we are doing with our fight against Ezra Woods' bill that was presented to the City Council as a way to delay our request or registration of our sales office in the zoning area."

Raphael called Tom back. He said to him, "Ezra Woods needs seven of the twelve City Council votes to pass the bill. We so far have our democratic friends, Stanley and Carmen, the two new dairymen who we have taken into the network. We have Harvey and the police vote, and I believe the chairman, Phillip, is on our side. He seems to favor what we're doing, but he is also a close friend of Joe Hendricks, who pushed Ezra into creating the bill in the first place. The northwest councilman is doubtful; he's often a no-show at the council meetings. We need to be sure of the northwestern vote at the council meeting next month."

Tom said, "So, we have five, and we need two more to be sure that we'll get our Bill passed, except if we have trouble with the council chairman."

Raphael said, "Yeah, that's all for now."

Stanley said, "If Jim Gurney doesn't vote to kill Ezra's bill, he'll pay the price with the Association. Our sales office can give our Association a nice steady income, so we don't have to depend entirely on memberships."

Henry said, "Stan and I will escort Jim Gurney to the council meeting to ensure he votes the way the Association wants him to."

Tom then said with a smile, "Okay, how do we get to the chairman?"

Abbey spoke up, "I've known Phil for many years. I feel he'll favor what we're doing and see that it is good for Cliffside and our carbonization program. Let me talk to him."

"Well, okay," Tom said, "I think we're in good shape."

Before they realized it, a couple of weeks had passed. It was soon six a.m. on a Sunday morning at the yacht parking lot.

Stanford appeared at the top of the stairs and said, "Follow me on board."

They all crowded down the steps, into the boathouse, and onto the yacht. The captain introduced himself, and his crew invited them to take a seat as the motors were already humming.

Out into the Bay, the captain turned east, and soon, they were facing the open Atlantic.

Stamford invited them all below deck, where breakfast was waiting. After a few moments, they were settled. Tom set his coffee cup down and said, "George, give us a program report on the sales panels and the development of their seven-day home emergency program—which I hope will turn out to be a real winner in our green power industry."

Priscilla spoke up, "Tom, first breakfast, then business, okay?"

Nancy said, "Sausages and half-hard-boiled eggs and hot, buttered biscuits."

She looked at George pleadingly. George smiled and waved her on.

The others took the hint and dug into the food. Tom said, "I apologize for my manners. Let's enjoy our trip."

George said after they had eaten, "The forty-five-acre array produces generally six MW hours on a sunny day. In the evening, we have our wind power coming in from Stamford's farm that's working out well, too. We are scheduled to change to a more efficient and exciting solar-powered chemistry. I'll fill you in on all of that when the time comes; it's still a little way down the road."

Tom interrupted and said, "Tell us about it now, we have time."

George said, "There is a new miracle material called 'perovskite' that was named after the scientist that discovered it some thirty years ago. Some scientists have been experimenting with it and call it a miracle material, and it has been tested to improve the sunlight-driven energy efficiency of solar panels. Let me repeat that: It's going to improve the efficiency of solar panels by mixing it with the chemicals we're using now. Perovskite is not exactly new; scientists have already known its potential to be used in solar panels. However, that potential had never been truly tested. I'm quoting from some scientific material; it's not my own research. Two separate studies and my teams of researchers from China and Japan published a paper on Thursday that have conclusively proven that perovskite can achieve the efficiency of photovoltaic solar cells by over thirty percent above the known limit of twenty-nine for traditional PV solar cells made with silicon.

"The test results have concluded that it's possible to push the threshold of the solar panels and produce low-cost solar panels for consumers. They have also discovered a new solar cell that relies on night vision technology to generate and store electricity even in complete darkness. It seems that solar energy is not on the rise by using silicon alone, but the newly developed perovskite structure is a compound that has the form AB X3, the same crystal structure as petrol perovskite. The easiest way to describe petrol perovskite's structure is as a cube, not a cell, with a calcium atom in the middle. Instead, it has oxygen atoms near the edges and titanium atoms at the corners.

"I hope you're all getting this technical information that Tom asked for. In the meantime, because of supply chain issues, the production of silicon solar panels has been put on the back burner. Chinese and Japanese scientists behind the latest breakthrough are planning to create new, more efficient, and cheaper solar panels. A startup in China has already announced a plan to start production of tandem solar cells with perovskite, cutting costs to only a hundred-twentieth of traditional solar cells. It is estimated by some that their solar panels produced fifty times more power than the standard panel used today.

"That's the technical answer to Tom's question, however, it's going to be a little while before we see any of that. I'll certainly keep you up to date on it. We must convert our forty-five acres to the new tandem mix, and we have to find a way to do it easily and economically."

Tom said, "I can see that it's not going to be a simple thing to do. When the time comes, we'll put all our heads together. I'm an electrical engineer, and George is an electric engineer, and Henry has many years of experience in the practical world, so we'll do the best we can."

George continued, "Morning Sun has set up a large room for two men who are experimenting with the old-style solar panel for our home security plan. That includes the panels, the turbines, the generators and an insulated thirty-gallon, thousand-degree energy sand barrel that would supply an average home in case of an emergency, but at the same time, could normally provide a twenty-four-hour daily power source. Their main problem now seems to be how to switch the system back off emergencies to the normal daily pattern. We can successfully switch to emergencies, but we can't find a way to switch back. We want the solar panels to keep producing, but we don't want the turbines to continue to run and generate power at the emergency rate."

Tom interrupted, "Okay, let's leave it there for now. Set up a Zoom meeting, and we'll all get together when we have an opportunity."

Tom then turned to Abbey and said, "You and I have already talked on security. Proceed with the plans that you told me about and keep me informed. But right now, please work on the City Council; we don't want our sales office to have a hang up."

CHAPTER ELEVEN

Phoebe cried out, "Look! We can see the eight towers in the distance."

Tom said, "Oh good, right on time–Stamford you've got the floor."

Stamford said, "Let's all stand up around the rear rails. Captain, slow down. I want to talk as we round the far towers."

Stamford continued, "We can see two rows of four towers a safe distance apart. But below the surface, they are all connected by cables for control and information. The tower with the attached building is our home and central work area—so each tower has a control room with access at its base.

"It all has been designed, built, and run under the supervision of the National Oceanographic and Atmospheric Administration within government. We use NOAA for short. They have developed a system to forecast marine changes, such as heat waves, based on thirty years of satellite data. They release monthly forecasts for the Atlantic and Pacific ocean.

"Among other things, the forecast shows this month that the marine heat waves are likely to linger through the rest of the year. They show the world's oceans have been warming and have absorbed an increasing amount of greenhouse gases. The temperature of the ocean expands and contracts over the North Atlantic by what is known as the Bermuda High, or the Azores High.

"The high pressure controls the trade winds. When weaker winds blow over the ocean, it lowers the temperature at the surface. The El Niño of the Pacific usually creates a vertical wind shear over the Atlantic. The warming Atlantic creates clouds, slowing the wind speeds, and forming a tug of war between the Atlantic and the Pacific.

"The temperature of the Atlantic is closely watched by our satellites. If it climbs to fifty degrees, when it is usually forty degrees, it's not considered a good sign. We then have to adjust to this information.

"The ocean warmth, of course, could muster a very hot summer. It does damage to sea birds and fish; their populations have declined seventy percent in some areas this last year. As you can imagine, it controls the wind as well, so we watch it very closely and control our demands accordingly.

"I see we have reached the tower with the attached building—let's get aboard our office structure, and we'll give you a tour of the tower itself."

They followed Stamford onto the deck of the building and up the stairs to the dining room and social room.

Stamford then said, "Follow me into the rest of the building."

They passed the kitchen then a soundproof bunk room with showers. Rafael asked, "Why the soundproof room?"

Stamford replied, "We have three shifts; someone is always sleeping and we don't want to disturb their rest, so we soundproofed the room from the rest of the tower and the kitchen. We also suggest no talking while in the bunk room."

The next room was a chart room with walls covered with charts of all types and descriptions.

"NOAA has the most prominent communication equipment," Stamford said. "We can talk to any place in the world, or to any ship at sea."

The last room he took them to was full of controls, broken down in eight tower sections. Stamford pointed each one out in turn and said they all had similar communication.

They then returned to the social and dining room. Stamford said, "Find yourself a comfortable chair. Dinner will be at five p.m. Excuse me, I have a couple of hours of work to do."

Once everyone was seated, Tom said, "I noticed the tables have snacks on them, and we're probably a little bit hungry, so help yourself."

Then he said, "Phoebe, what can you tell us about the classrooms on the first and second floor of our network building?"

Phoebe answered, "On the first floor to the right, Peter Whitehorse has set up his architectural computer drawing department and a connection to their Boston office printing department. But that could easily be converted into our boardroom for meetings. The room across the hall is set up as a public visitor welcoming room with a receptionist, comfortable chairs, and a switchboard to each of our departments. The two back rooms will be for teaching licensed public electricians and distribution linemen, which will be supervised by our general manager, Henry.

"The four rooms on the second floor are for solar panel power education, hydroelectric introduction, Agrivoltaic, using robots, and new development of solar panels.

"I'm working with John Stevenson, former Board of Education member and the superintendent of schools. They are interested in our vocational curriculum and true electric power. However, there's been no offer of financial support, yet."

Henry asked, "Do you have any instructors in mind? I've got four qualified electricians working for us now; maybe they could act as downtime instructors."

Tom interjected, "If they can do the job, it would save a lot of money. I'm in favor of the idea."

Phoebe added, "If they were willing to take a training course at our trade school, that could work."

Henry added, "I'll discuss it with them and let you know."

Tom added, "If anyone has any comments about the reports that we've brought up to this point, speak up now."

No one spoke, so Tom asked Nancy, "Have you got it all?'

Nancy replied, "I've recorded every word since we left Cliffside."

Tom replied, "Good, make sure everybody gets a typed copy."

They all gathered in the social room, resting and admiring everything that they'd done out at sea. It certainly was a scientific marvel—a community of family, modern men and women, and sailors without ships. The conversation was imperative to the initiative to save our planet from encroaching carbonation.

The time passed quickly, and then there came an odor from the kitchen. Roast beef, corn, peas, onions, and baked sweet potatoes. All followed by a cream of potato soup topped with bacon chips.

Priscilla commented, "Wow, that couldn't have been better in New York City."

The young men serving it laughed and commented, "We love having you here, but I must admit, we don't eat like this every day. Wait till you see the dessert—banana cream pie."

No one talked for an hour when the dessert was served.

Tom got up and said, "I suggest that we move to have the Green Power Network give each member of the wind farm staff a fifty-dollar bonus— all in favor say *aye*."

They all shouted *aye*. Stamford got up and said, "Thank you all very much, we love having you here."

They quietly sailed for home, all sitting on the rear deck with just one lamp—it was very romantic. Stamford, who was sitting next to Priscilla, said quietly, "We have quite a family of couples. Abbey and Phoebe in their seventies, you and I in our fifties, George and Nancy in their twenties. It would be nice if we invited our wives next time—Raphael and his wife included."

Priscilla tapped Stamford's knee and said, "They all make up the next generation. They'll deal with climate change together."

The hum of the yacht motored over fifty miles of the Atlantic. In two hours, they were together back in the Cliffside Bay.

CHAPTER TWELVE

Joe Hendricks was sitting in his office at his main gas station when his door opened. A tall, dark haired white man entered and asked, "Are you Joe Hendricks, the owner of six service stations here in Cliffside?"

Joe looked up over his desk with suspicion. "Yes. How can I help you?"

The man looked suspiciously around and then said, "I'm Kirk Forest. I've been sent here to assist you in solving your green power problem."

Joe asked quietly, "Who sent you?"

"Let's just say I'm from a dark money program; I'm a specialist at solving green power problems. Just show me where they are, and I'll investigate and report back to you with possible solutions. If you approve of my solutions, I'll do the job and disappear."

Joe got up from his desk, walked around to a cabinet, and pulled out a map of the city and the area around Cliffside. He pointed out the locations of the Carlton farm, the array of forty-five acres of solar power, and their main plant.

Kirk said, "Good, give me a week. I'll check them out and make some suggestions before I do anything."

He opened the door and left with a map and drove away in an old Ford pickup truck.

Neither of them knew of Eclipse, the police undercover group originally formed by now-retired Chief Detective Abediah Tutt. It had been developed twenty years ago to find Joe McCarthy-type communist infiltrators. They had planted long-range microphones all over Cliffside, including in Joe Hendrik's stations.

A pleasant young woman, sitting in the old coal factory basement of the Eclipse Network, was in a room hidden behind the snack bar, smiling as she recorded their conversation. Abbey's newly revived Green Power Network Eclipse security group was worth it.

When Abbey heard the news and read the report, he urged the group to watch all observers who were curious about who they were and what they did.

Over the next several days, Abbey got several reports from people and the police, all mentioning a Ford pickup with out-of-state plates.

He then asked the police chief to send a national inquiry on the Ford pickup and the name Kirk Forest.

After that, Abbey shared the pickup's plate with Tom and Henry. Tom said, "I'll tell the lieutenant to tell his people they did a good job. We'll run with our regular business."

Then he said to Henry, "On other business, I've been watching the recorded grid and the other businesses in the use of the sand heated silo. We're only using about fifty percent of the capacity at a thousand degrees. Let's develop a heat program or baseboard radiation. We'll heat our building, then the sales offices at the association, the farmhouse, the business building, and the quarters at the array.

"Hire the most experienced installers of baseboard radiation heating and plumbing. I'll ask Peter to have his architects create working drawings for those buildings.

"We won't use public water in them. We'll use the water from our pond to maintain the heat radiation. We built that pond many years ago at the end of the plateau to assist our irrigation in the dry season. It's ten feet deep at the dam wall. I'll have him put our supply tax at the bottom of the deepest end and make sure it's out of sight. Keep it to yourself for now. I'll tell Peter to include some instructions with the drawings.

"It will be a downhill flow all the way from our pond. We used to use it as a swimming hole in August, but it was too cold early in the season. Otherwise it's been our farm's underground well support for all these years."

Henry said, "I was telling him about the drawings. He'll check to make sure everything is up to date and replace parts if necessary. We'll keep it a secret. We'll use our underground power routes and tell everyone we are doing a service check to make sure everything is okay."

Tom said, "I think it's a great idea. We might be able to create heat for some of the northwest residents' homes once we've determined its capacity—including the farmhouse. We won't dismantle the heat we're using but will improve the baseboard radiator with the original Colonial design to preserve its historical value. The farm's heat has, for years, come from wood-burning stoves since the Colonial days. The family tradition has always been to keep it intact for historical reasons. It was always a highlight on the Fourth of July town when they toured historical homes.

"We have always used wood in the kitchen stove, the fireplace in the living room, and a potbelly stove that heats the upstairs room through a vent in the ceiling and a stove pipe. We have always banked the fires at night with coal to keep them heating all night long. My mother always urged my father, Tom, to put in a coal fire furnace, but he never agreed to it.

"I told Peter that if they have to change the boarding anywhere, to use natural wood colors in the store and the smaller barn. I also told him we wouldn't need heat in the large barn because the herd of cows naturally keeps it warm in winter. The building is controlled already through a natural generator with electric heat, so we don't need to worry about that."

One day, when Tom was having lunch with the family and George, a farmhand named Jeff approached him. He came up to Tom and asked, "Would you explain some things to me?"

Tom said, "Sure, what's on your mind?"

Jeff said, "I'm trying to explain Agrivoltaic to my family, and they understand the use of fourteen-foot-high solar panels, and how they help vegetables grow by preserving water. What we do not understand is the use of the coal bin in the main building that's full of hot sand."

Tom said, "We need a place to keep the excess electric power that we don't need right away. We can keep the energy in good condition in hot sand until we need it."

Tom continued by asking, "When we're on the farm and our stock needs food in winter, where do we get it?"

Jeff answered, "From the silo and haylofts."

Tom answered, "You're right. And that's exactly what we're doing with the sand—preserving our power until we need it."

Jeff then said, "Why don't you call it a silo? If you want to sell your Agrivoltaic to farmers, talk to them in a language they understand."

Tom answered, smiling, "You're right Jeff—we'll call it the silo from now on. Thank you."

Jeff answered, "Just trying to help," and he walked away.

Tom picked up his smartphone and called Henry and told him about him and Jeff's conversation. Henry said, "Good idea, I'll put a sign on the east and west side walls right away."

Tom then called George and told him the story. He also agreed saying, "It is a good sales suggestion."

Back in the espionage room behind the basement snack bar in the main building, the secret agent was sitting in her seat when a field microphone receiver clicked on. She put on her headphones and turned on the recorder.

Kirk Forest's voice came through loud and clear. "I'm back, Joe, and I have a couple of program ideas that will certainly shut down our competitive green power operation, at least for a little while."

Joe asked, "What are they?"

Kirk answered, "The idea that would do the most damage is blowing up the roaring brook dam, just off the array at the farm. The water roaring down the brook bed will certainly take out the coal building that they're using as a headquarters."

Joe said, "Whoa, no way. That dam has enough water in it to wipe out half of the homes in the northwest of Cliffside—all the way down through the waterfront. Many of those people are customers of mine and have been for a long time."

Forest seemed disappointed but said, "Okay, we won't blow up the dam."

He continued, "The next best idea I have is a night flight in a helicopter over the plateau full of dry hay. We'll drop the hay over the backside of the array, all over the vegetables, and on the plateau itself. Then we'll swing back and drop burning torches on the array and half of the foliage. It'll start a great fire."

Joe said, "I want to wipe out that forty-five-acre array, but that threatens the whole farm, so unless the fire department can stop it in time… What's your next idea?"

Kirk said, "The main office building sucks water from the brook, which passes by the south side of the building. They convert the water to hydroelectric power to sell to the grid. We could easily slip a dynamite charge into the brook, at the proper point. It'll go inside the water flow and explode inside the building."

"Wow," said Joe, "I like that. What else is on your list?"

Kirk said, "We can easily set the farmstand and barns on fire with rockets from far up on the ridge."

Joe said, "I don't want to kill the people. Okay, burn the farmstand. But don't touch their Colonial homestead. Our town is very proud of it, and we don't want to lose it for our Fourth of July celebration. Boy, you national fossil fuel troubleshooters play rough."

"Okay," Kirk said. "We'll preserve the Colonial building for the town. Yes, we do play rough; we have millions invested in fossil fuels and we don't intend to lose it to entrepreneurs all over the country. How about I come back in another couple of days? I'll get set up to burn the array, the barns on the plateau, and blow up their main plant from the brook. Then, I'll arrange for my disappearance."

The lady behind the snack bar copied and reported the whole conversation to Abbey.

The day of decision with the City Council came. At the meeting, Raphael Sullivan, Phoebe Sinclair, Abbey, Henry, and Stanley, who was their charge from the Northwest Residents' Association (which had a vote on City Council). They all could be found in their seats, waiting for the meeting to begin.

The first hour of business was the usual Cliffside affairs and the suggestions for the new mayor.

The chairman then called for new business. Ezra Woods raised his hand for permission to speak. The chairman acknowledged and recognized him. He stood up and read the bill he proposed, "My bill points out that the Green Power Network's request for a solar power sales office would give them a zoning permit to sell solar power to businesses and private homes. My bill also suggests a period of study to see its effect on the community prior to being approved. We feel that it would have a serious negative effect on the existing businesses in the community."

The chairman hit his gavel, gave the bill a number, and asked, "Does any council member have any questions?"

Carmen Spinoza raised her hand. She was recognized by the chairman and she spoke eloquently on the need for climate control in their community and the value of the Green Power Network helping to protect the people through emergencies such as fire, flood, or hurricane, when the existing power grid was overloaded. In addition, unlike fossil fuels, they did not add carbon to the atmosphere.

Another member's hand came up from the Dairymen's Association. He spoke about the financial benefit their members have had from joining the Green Power Network. He explained that it helped a shaky market by giving them an additional source of revenue from the sale of electric power back to the grid.

A fossil fuel merchant member stood up and claimed there to be serious damage to other existing merchants because of cheaper green power. He demanded it be studied for a while.

Finally, the chairman called for a secret ballot. The ballots were distributed to each member. No comments were allowed. The ballots were then collected. After they voted, the ballots were counted, and the results were handed to the chairman. He read them carefully and announced the vote was seven to five—in favor of awarding the zoning permit to the Green Power Network.

The news was greeted with great applause from the gallery, as well as most of the members.

Raphael called Tom on his smartphone to give him the good news.

Tom then told Raphael to tell Stanley to open the sales office at the Association as soon as humanly possible.

Abbey nudged Phoebe and urged her to watch Joe Hendricks, who had picked up his private phone and was calling someone urgently. Phoebe responded, "I'll wager that isn't good news for us."

Raphael, Abbey, and Phoebe congratulated the City Council members who voted in favor of the Green Power Network and the zoning permit for the public sales office.

The evening was a success, and they all felt better about it. On the way back to the Painted Lady, Abbey told his agents about the new developments and suggested they step up security efforts targeting people who were not employees of the Green Power Network. Including people like Patrick McCormick or Bill O'Donnell of the Farmers' Association. The Police Department would also get a full copy of Forest's threats.

Abbey suggested a light snack at the Painted Lady to run over the details and some new ideas. Phoebe said, with a knowing smile, "I wondered when we would get around to getting together again."

Abbey smiled. "We'll get to that with the coffee. I've missed it myself," he said as he tapped her gently on the knee.

"First, though, from now on, you and I will wear our shoulder holsters—Hendricks' people may play rough. You and I want to be prepared."

Abbey put on the coffee when they arrived and took out the cupcakes. They sat close on the divan and talked about their vulnerabilities and their partners in the Green Power Network.

Abbey told Phoebe how he received the message from his agent behind the snack bar about Kirk Forest's ideas. The microphones, which were secretly placed, recorded how he might cause potential floods and fire burnings and bombs on the farm, which proved there was no doubt about Joe Hendricks' determination to stop green power expansion. He knew Tom Carlton would be concerned about his family, his neighbors,

George, and the Morning Sun staff who were working every day on the power array.

Abbey looked seriously at Phoebe and said, "Our people have been successful doing what they wanted because they were on home ground, doing what they know, with neighbors who want to see them succeed. Now, they are open to the public, who may not care where they come from. I'm sure they have attracted the attention of the big fossil fuel companies with big bucks and a lack of personal feelings.

"We must face the forefront now—you and I—because of our training and connections with those who Protect and Serve. You and I know what is coming, and it is going to be beyond our folks' abilities."

Phoebe said, "We're the cavalry that's supposed to show up in the nick of time—what's our first move?"

Abbey laughed, leaned heavily over her, and said, "Just make sure you're prepared" as he kissed her.

She giggled. "Are you sure your gun is loaded?"

He said, "I never put an unloaded gun in my holster."

Phoebe squeezed him hard and said, "I'm counting on that."

Breakfast in the Painted Lady dining room was sunlit and pleasant. They had two eggs, hash-browns, pork sausages, and, to restore their energy, coffee and large jelly doughnuts.

They sat back, and Abbey called the chief and made an appointment. They visited the chief and explained the entire situation, including the threats suggested by Kirk Forest.

Phoebe said, "The Carltons are good people, and their son, Tom, Jr., is a graduate of electrical engineering with a master's degree in solar green power. He believes in replacing fossil fuel power because of its huge carbon poisoning of our environment."

Abbey inserted, "But they need professional protection from the dark money subversives paid for by the fossil fuel establishments that want to stop the Green Power Network development. Chief, I, as a retired chief of detectives for this department, along with Phoebe—who is still on the police pistol team and an auxiliary policewoman—we ask you to create a trained force to back us up with investigative skills and force, if necessary, for our safety in preserving the sales and services of solar power in the Cliffside program. Can you help us?"

The chief looked hard at them and asked, "Tell me about your new Green Power Network."

Abbey replied, "We operate a forty-five-acre solar panel array on the Carlton farm in Northwest Cliffside, which produces megawatts of electrical power each hour when the sun shines. An eight-tower wind farm off the coast in the Atlantic is also involved and provides power for us at night. Both forms of energy are received and forwarded to the Cliffside grid through the converted coal plant—our home office on Factory Place in Northwest Cliffside—we call it the Green Power Network.

"We're opening a public sales office in the Northwest Residents' Association building shortly, and we have an agreement with the Cliffside grid that allows, in the case of public climate related emergencies, to cease funding power to them directly from the grid and to use it as an emergency power for police, fire, and medical damage for citizens who need help from EMTs, the Red Cross, or the Salvation Army.

"Farmers or dairy-workers who have our services also benefit from an annual cash payment, which comes from our excess power that's maintained in the heated sand silo storage."

The chief turned and asked his secretary if she had recorded all of that. She replied, "Yes, I have."

The chief said, "We've been watching and collecting information on you for some time. We have waited for you to ask us to get involved.

"We owe your corporation, Forever Blue Public Trust, and will help.

Yes, we'll create the task force as soon as we can to keep your people and your network safe from outside interference of any kind."

On the way over to the Green Power Plant, Abbey told Phoebe that he was feeling a lot better now that the chief had agreed to create a task force to protect the Green Power Network.

Phoebe answered, "Me, too. Now I feel Tom can proceed with his plan for emergency support. His first step, I hear, will be strengthening out-of-town EMT units in the east and west counties as well as finding a way to help the Red Cross and the Salvation Army during an emergency."

Abbey added, "The other thing we have to publicize is our emergency number, though I am worried about being overwhelmed with calls for help. Of course, that's assuming the state legislature will put several support team members and their equipment on a back burner when they vote on the annual budget.

"Cliffside emergency numbers have been put in place at a fixed fee. If they drop the ball, our Cliffside citizens will be in trouble. In the past, scammers have rushed in to take advantage of the desperate people who need help. Unless the Red Cross or Salvation Army are right there to step in—which is why we have them on our list for support.

"The climate control experts claim that droughts, tornadoes, and heavy rains will be happening more often this year. The decarbonization of our atmosphere is ignored by fossil fuel interests, and the rest of us seem too busy to care. Most of the public doesn't seem to realize that the large industrial smokestacks continue to pump out carbon every day. Since we don't appear to be able to control the fossil fuel profiteers, those of us who have the ability must do what we can to be prepared to survive climate change."

CHAPTER THIRTEEN

Phoebe said, "Please drive to the farmstand; I want to buy some fresh fruit."

Abbey replied, "Okay."

Phoebe persisted and said, "And you need more half-and-half for your coffee in the apartment. Plus I'm anxious to see the new dairy part of the farmstand."

Abbey parked the car in the east parking lot of the farmstand and entered the shop. Abbey stopped, turned, and took another look at the parked grey four-door Buick sedan. He continued inside and looked around then quickly walked over to a slim man and asked, "Jed, why are you here at the farmstand?"

Jedediah Tutt spun around and said, "Hi Dad, what brings you out in the daylight?"

Abbey laughed and said, "Phoebe wanted some fresh fruit and to see the new Carlton diary wing."

Nancy, who was behind the counter, heard all this and came around quickly.

She said, "Jed, I didn't know you are Abbey's son—does that mean Gerry who was in here earlier is too?"

Jed answered, "We've been your customers for a long time, but the subject just never came up."

Phoebe came over and gave Jed a big hug.

"Stay here," Nancy said. "Mom has to see this."

She came back with Priscilla, who said, "Talk about a family get together. Phoebe and I go back to her high school teaching days—and Abbey snuck himself in there somehow. I guess we've all joined together in Tom, Jr.'s new Green Power Network."

Jed said, "I've noticed Tom's new monster solar panel array out there on the plateau. I didn't know he had taken over the old coal plant."

Nancy said, "That's our new corporate headquarters."

Abbey added, "I'll take you both through it, if you have the time."

Jed answered, "I think Gerry would like to see it, too."

Abbey responded, "Call him as soon as you can and invite him. We'd like to see him, and we'll take him on a tour."

Jed answered, "I'll call him when he gets home from work, bring him up to date, and call you after supper at the Painted Lady."

Priscilla asked, "Who's the Painted Lady?"

Phoebe laughed and said, "Pris, that's what they call our senior housing Victorian mansion."

Priscilla answered, "Pretty classy."

Abbey added, "We'll take you both through the array—George Ralston is the electrical engineer and supervisor of the array here on the farm— then we'll go down to the headquarters. You'll get a chance to meet Tom, and he'll give you a three-hour lecture on the future of green power and Cliffside."

Priscilla laughed. "Don't get him started."

Jed replied, "I've got to run now, but I'll call you later, Dad," and he left for his car.

George had come in while they were shopping, and Nancy brought him up to date. He was surprised to learn Abbey had two grown sons.

Nancy said, "Jed's the millionaire owner of Cliffside Maple Products Corporation, and his brother Gerry is the vice president of the railroad in Bayside."

George said to Abbey, "You surely know how to keep a secret. How come there aren't any railroads in Cliffside?"

Abbey replied, "Gerry's been a railroader since I bought him his first one at age eight. You'll have one in Cliffside one of these days if he has anything to do with it. Tomorrow we will get a chance to talk to both of them when we give them the grand tour."

Phoebe finished shopping, including the half-and-half for Abbey's coffee. They said their goodbyes and left for Abbey's car.

True to his word, Jed called his dad after supper and said that Gerry was looking forward to the tour of Tom Carlton's new Green Power Network. He said that Gerry had heard about it and was in favor of non-fossil fuel sources of electric power instead of the diesel fuel used in the railroads now.

Abbey, Phoebe, George, Nancy, Priscilla, Jed, and Gerry were all together enjoying Priscilla's coffee on the farm the next morning.

Jed and Gerry were full of questions about green power. Jed's first question was: "Why do you have four charging lanes for electric vehicles?"

Nancy answered, "We have our own chargers for electric power because we have electric vehicles on the farm, as well as our personal cars. We maintain the four charging stations in our parking lot for our customers. There's a significant benefit. We sell power through the grid's installation in Cliffside. We produce six megawatts an hour from our

Agrivoltaic array. I guess it's time to start our tour, so let's walk across the field to the control building."

When they arrived, George explained the basics of construction.

"We have forty-five acres of solar panels mounted on steel frames fourteen feet above the ground, and each row is twenty feet apart. The panels above catch the sun's rays, and they produce an electric direct current, which is converted to alternating current—that is the power you use in your home. We send that current through a transformer to the Cliffside grid to sell through their distribution to homeowners.

"The Green Power Network does not create any carbon like fossil fuels do, and, of course, it does not pollute the atmosphere. Since the Carltons own this solar power installation, they get first dibs on the power they use to run their farm, their farm's market, their home's electric appliances, their own heat, and to run their personal vehicles."

Jed turned to George and asked, "Can you put my plant and home on your Green Power Network with two charging stations? One for the plant, and one by the garage for my home?"

George asked, "Jed, where is your home and plant?"

Jed answered, "On this road, just east of here by three-fourths of a mile. My home is right next to my plant."

George answered, "I can run an overhead line down on the poles, right through your plant from here."

Jed said, "Sounds good."

Abbey added, "Pat McCormick is converting our vehicles to electric. If you want, he can change yours, too."

Jed reached over, shook hands with George, and said, "We got a deal—call me when you're ready to do it."

Jed then turned to Abbey and asked, "I know you're trying to move your solar power to the western counties—do you suppose my maple sugar suppliers might be interested?"

Abbey answered, "Why not? Why don't you ask them? If they're interested, I'll have Stanley, our sales manager, meet with them and give them the details. They could meet anytime at the plant. If they have a farm, they may want to see the Agrivoltaic installation right here."

While all this deal-making was going on, they were all walking under the solar panels and enjoying the water-saving shade as the panels turned, following the sun.

Gerry asked, "What happens after dark?"

George answered, "The panels rest, and the eight-tower wind farm fifty miles out in the Atlantic sends us their nighttime power, which we store in our sand silo, or we send it on through the plant. It's a twenty-four hour a day operation."

Priscilla added, "It also provides a nice winter income for our family."

Gerry asked George, "Are you coming with us down to the plant?"

George answered, "Thank you, but my work is here, taking care of my solar panels. I'll see you again shortly, though."

Tom was waiting for them in the lobby. Priscilla, as the corporation's president, stepped up to talk to Jed and Gerry. She said to Tom, "I don't know if you remember them, Tom. These gentlemen are Jed and Gerry—you have helped them many times at the farmstand before you went off to school. As it turned out, they're the sons of our friend Abbey."

Tom replied, "Now that you mention it, I do remember them, Mom. They were farmstand customers for many years. It's good to know they're Abbey's sons."

He shook hands with both of them and said, "Abbey must be proud of both of you."

Jed said, "Well, looking around here, you sure have hit the ground running since you finished school."

Tom confessed, "Your dad and my uncle Stamford, who are officers with Phoebe at the Forever Blue Public Trust, arranged for the purchase of this building, its renovation, and the solar panels in the Agrivoltaic installation. Without their financial banking, I couldn't have come anywhere near this far this quickly."

Abbey spoke up and said, "It's all we could do to keep up with your entrepreneurial drive."

Tom flushed a little and said, "Let me introduce some of our key staff. Here is Henry Lipton, our general manager and the one with the most practical electrical background; he was a foreman of the linemen with our local power company before he retired.

"The gentleman beside him is Stanley Kowalski, our sales manager. Both men are neighbors in the Northwest Residents' Association. The man right next to Stanley is Peter Whitehorse, our chief architect from Wickiup in Boston, and a graduate from the Algonquin University on the Abenaki Reservation, just north of Cliffside."

Peter said, "His ideas keep us happy and working all the time."

Priscilla spoke up again and said, "We came for a tour—are you prepared Tom?"

Tom, Henry, Stanley, and Peter took turns explaining what they did to produce a green, non-carbon, electric power and how they sold it to the Cliffside grid substation. They traveled through the plant for two hours.

They ended up at the snack bar on the ground floor for lunch. They ate sandwiches and made small talk. Tom went into his further mission—to assist Cliffside's citizens during national emergencies, like a hurricane or a flood.

"Our Green Power Network will stop production for the grid and start producing emergency help for those persons and organizations that are usually the last being considered by the city budget. Including the Red Cross, the Salvation Army, the Police Department, the Fire Department, medical locations like hospitals, and of course, the EMTs and other people outside of our city limits. It seems that whenever an emergency arrives, because they are not city-supported, they don't get considered right away."

Abbey added, "We're proud to provide carbon free assistance in our battle for climate control."

Jed said, "You have a right to be proud. I'll communicate with my customers that the Green Power Network is here to help you and your farm."

They all finished lunch and prepared to return to their daily duties.

CHAPTER FOURTEEN

After walking Jed and Jerry to their cars and thanking them for their support, Tom went back upstairs to his office. He was surprised to find Henry at the window using a pair of binoculars. He asked, "What's so interesting?"

Henry answered, "Two guys under the Hill Street bridge."

Tim responded, "Remember the police's secret microphone. The guide who calls himself Kirk Forest mentioned the Hill Street bridge and that it was close to our plant. He told Hendricks it was a potential target to shut down the plant."

Henry answered, "Yes, it was the number one target—to blow up the roaring brook down on the farm."

Tom said, "Yes, but Joe Hendricks said *no thank you* to that because the community was fond of it and the water would come rushing down, destroying homes and businesses all the way down to the waterfront. He doesn't want that to happen."

Henry continued, "I'm looking carefully at these guys, and I don't know the dark-haired one, but I think I know the bald one. Look here through my binoculars and see what you think."

Tom took the binoculars and studied the guys moving in and out from under the bridge. He asked, "I wonder what they find interesting? I think the bald guy works for Joe Hendricks. The dark-headed guy is probably that one they called Kirk."

Henry said, "Look down Hill Street. What do you see there?"

Tom changed his view and said, "There's the black Chevy sedan our police lieutenant drives. Abbey's undercover crew are on the job—he is a good man, our retired chief of detectives."

About two hours later, Abbey came up to Tom's office and asked him to ask Henry and Stanley to join them.

They arrived with Abbey, and met the lieutenant, who also gave the same report about the two men under the Hill Street Bridge. The police had trailed both the men back to Joe's gas station and then trailed Kirk to an out-of-town motel. They took pictures of all this and his Ford pickup with the out-of-state plates.

Tom then told Abbey and Stanley about him and Henry's observations of Kirk and the bald guy. The lieutenant had reported that the bald guy was Walter Ciccone, a longtime employee of Joe Hendricks. The police had pictures, addresses, and home phone numbers for both the men.

Tom turned to Stanley and asked, "Were you here a long time before the steel bridge was built? Is there anything special underneath it? The two men keep going under there and popping back up again."

Stanley sat down and thought a long moment then said, "The only thing I recall is that one of the men, our neighbor at the time, said that the coal company management asked them to bank the bottom of the brook to make it high on the far side and low on the building side so that the depth would cause the water to roll from the bridge, close to the plant."

Tom asked out loud, "Why would they want the water rushing faster towards the building?"

Abbey smiled and said, "Come over here, you guys, you're too young to remember the hard cold days. Then, they needed additional water pressure to wash the coal dust from the hard coal to get it into the bin so it would burn slowly and hotter. The dust probably went downstream."

Henry then said, "That would be illegal today; it's pollution of a waterway."

Henry continued, "We have made great changes in the design of the water as it enters the building and into the basement. The new steel system filters clean water and rushes all else back into the brook."

Tom added, "One cannot see the new steel system since we covered it on the outside with red bricks that cover the rest of the building through the hole, but the water that comes in is still visible."

Henry said, "Kirk thinks they can send an explosive or a projectile of some kind into the interior of the plant building through that opening, and maybe blow us up."

Tom added, "Sounds like a logistical mistake on their part. I suggest we keep what we know about our hydroelectric system within the building and the main generators. As a matter of fact, all of that stuff on the main floor is off limit to visitors from now on. In any case, we'll hope the police discover them before they put their plan into action."

Stanley said, "Hold on, guys—Peter and his men designed our water system from the pond to all the way down here, including that part outside the brook. So, let's let him in on all of this to make sure his steel hydro structures won't be affected by anything that we do— already, parts of it might get damaged."

Tom answered, "Good thinking, Stanley. Safety first."

He picked up the phone and called Peter right away then said, "Abbey, on second thought, bring the police chief himself up to date and let him decide what he passes on. We want them all arrested for creating a criminal act."

On the phone, Peter reported to Tom that they received a warning from the National Weather Bureau; they were predicting a series of strong thunderstorms moving east from New York. They probably would reach them by tomorrow or the next day.

Tom replied, "I will shut down our hydroelectric generating power for a couple of days since the brook will be running high."

Tom repeated Peter's warning from the National Weather Bureau and the fact that it might reach them by tomorrow or the next day.

CHAPTER FIFTEEN

When Peter arrived, they reviewed the whole thing to him, and he said he would get with his people and go over it with a fine-tooth comb to make sure there wouldn't be any damage in what they were thinking.

Stanley said, "Some good news. I can happily report the finalization of six solar panel resident sales, with two more pending. We're waiting for the demonstration setups to arrive from Morning Sun along with their training people. They will subsequently train all the people once they have signed the papers and paid their deposit. We have selected the first purchaser as his demonstration home."

They all applauded. Tom added, "As soon as we can get their sales contracts and down payments, we'll issue the publicity bulletin to all of our employees, as well as to the financial supporters of our efforts, and issue a nice commission check to the Residents' Association."

Henry said, "The Association will be proud to be Green Power Network's first sales agency."

Tom picked up the phone and called Patrick McCormick Truck Sales. When Pat answered, Tom said, "Pat, I want you to order two fully equipped, electric EMT trucks, one for our east county and one for west County. They must include training for each group before delivery. We'll pay a third of the cost as a down payment and twelve-monthly payments to clean up the balance."

McCormick asked, "How soon do you want delivery?"

Tom said, "As soon as possible."

McCormick said, "I think I can do it in two or three weeks, if that works."

"Sounds good." Tom hung up the phone.

He told Henry, "When the new trucks are ready, let's arrange to pick up the old trucks they're using now for an electric conversion and for a complete checkup and repairs. When that's complete, include a nice new paint job for the old trucks. Our Green Power Network will pick up the tab for the work on the trucks as well.

"Peter, I want you to design two EMT buildings to hold two trucks each, one for the east, and one for the west counties. Each building needs charging units with solar panels. George will install the solar panels just like it would be on a home so that they can make their own power in case of an emergency."

Henry said, "All of this can be covered by EMT public service leases. We'll have Raphael draw up special contracts for each county. They will pay a hundred a month and the cost of their medical equipment. Anything we do for any of the groups—the Red Cross, the Salvation Army, police, fire—will be covered with these public service leases. If you have any questions or if they have any questions when this is happening, just refer them to Raphael, our attorney, and he'll work it out."

Tom said, "Terrific idea, Henry. I'll explain to Raphael what is wanted and that we are going to use public service leases as we need. We'll have each one of them sign up for it when we give them the trucks or any other equipment."

Tom picked up the phone and explained it all to Raphael.

Raphael asked, "Can you afford this kind of philanthropy?

Tom replied, "It's one our community will need in a climate change emergency. If you consider the public costs of calamities, like the fires on the Hawaiian island, Maui, or the fires in western states like California, or the floods in the Continental Southeast, then our Green Power Network preventive in Cliffside is to protect our citizens."

Tom added, "Let's build in some limitation clauses for our best protection. We already have an agreement with the grid that we will cease sending them power and switch to the support of whomever has our public service leases."

Rafael said, "Okay, I'll get back to you with some rough drafts for your approval."

Tom then turned to Stanley and asked, "Could you advise your associate member—the one who owns the Cliffside medical supply company—about our public service leases? Ask him to move his company up on the hill next to the hospital; he's in the flood zone with the river right across the street. Our Green Power Network will be willing to share the cost of the move as part of our emergency public service program."

Stanley answered, "I have no idea what he may think about it. But I'll raise the subject with him and see what he has to say."

The next morning, Tom was awakened by a clap of thunder. He got up out of bed, looked out the window, and saw it was raining heavily on the farm and the solar array to the west.

He went down to the kitchen to find his mother already preparing their breakfast. He commented, "It's raining pretty hard out there, isn't it?"

His mother answered, "The weatherman on the TV says there are a string of these thunderstorms coming from the west to the east."

"I wonder if the people of Cliffside are prepared?"

She answered, "They don't even think about it; they just take an umbrella and go to work."

Tom continued, "If it continues to rain for a couple of days, I wonder if the Police and Fire Departments are ready to help people. The people who live in the floodplain by the river to the west certainly will need support."

She said, "I think the Police and the Fire Departments have both gotten the weather forecast."

Tom answered, as he began to eat the bacon and eggs she had provided, "I'll call Abbey; he'll know."

His mother answered, "Not at this hour—let him sleep."

Tom quipped, "Who said let them sleep?"

She turned, smiled, and said, "You're getting too grown up."

Tom laughed and said, "I'll call him after nine a.m."

Nancy, coming into the kitchen, said, "The rain woke me up. What's so funny?"

Their mother said, "Tom was commenting on the storm, worrying about George out there with all that electric power and the rain."

Tom answered, "It's all insulated, Mom, and he's an electric engineer—an expert in his field. You don't need to worry about him."

Nancy added, "I worry too, Mom."

Tom laughed. "You're just worrying about losing your potential husband."

Nancy walked over to Tom, kissed him on the top of the head, and said, "You got it right brother, that's my dream."

Tom answered, "That's a good dream; he'd make a great brother-in-law."

Their mom added, "And a very nice son-in-law. I hope I like his family."

Tom looked out the window again and said, "That's some pretty heavy rain. I hope it doesn't last too long."

Their mother said, "The TV weatherman says the series is supposed to last a couple of days with a short period of sun in between. I expect it'll be slow at the farmstand for the next couple of days."

His mother came up, put her arms around Tom's shoulders, and said, "I remember you as a Boy Scout getting ready for a hike. You took your sweet time and took the Boy Scout motto, 'be prepared,' very seriously."

Tom gave her a squeeze and said, "Well, I guess Abbey is up by now." He went for the telephone.

Tom called Abbey and wasn't surprised when he picked up right away.

Tom asked, "How many rescue boats do the Police and the Fire Departments have? I'm concerned; the TV is reporting a couple of days of hard rain."

Abbey responded, "Our minds are in synch. The police have two old rubber boats gathering dust in the warehouse, and the Fire Department is the same. I know the police haven't used theirs in at least ten years. I'll call the chief."

Abbey called the chief of police and told him, "The Green Power Network is buying six rubber rescue boats with safety gear for the police and fire departments under our lease as part of the preparedness program. They're also getting the Coast Guard to come here and train our men in their use."

The chief was glad to hear it.

"I'll call the fire chief and tell them it's on the way so that they'll be prepared to receive those boats and to do the training."

Abbey said, "I'll call the boat company and make it a rush order. Also, I talked some time ago to the Coast Guard Academy on the Thames in Connecticut, and I'll give them a friendly call, too. When you talk to the PR person at the Coast Guard Academy, remind them of their offer to train our rescue boats' people. I told him there would be no cost to the Coast Guard and as a nice gift to their rescue boat training program."

After he'd hung up with Abbey, Tom then called the National Weather Bureau for some up-to-date facts on the cities that were experiencing the storms right now.

They reported that Upstate New York and Vermont would be impacted in the last days of the storm, which they expected to hit in three days. The storms were dropping three to six inches of rain each day. People in low-lying areas were already experiencing floods.

Tom called Abbey back and told him what the weather station had reported.

"Hopefully," Tom said, "it won't be as bad here, because they said the southernmost counties of northern New York and Vermont will be the worst hit."

However, after two days of continuous rain, the Bay had come up a whole foot alongside Main Street and was under six inches of water. The crossing of the bayside to Cliffside Bridge was threatened as it was now covered with a small layer of water.

On the Cliffside side of the Bay, the water was creeping up the waterfront side streets. The river coming in from the west was slightly over the banks and the water was in the cellars of those floodplain houses. Most of the Cliffside homes were on a side hill, but it was the floodplain that was the difficult area. Most of that was west of the city. The police station itself was two blocks above the waterfront, and the firehouse was three blocks above the waterfront east of the Police Department. However, the police parking lot, to the rear of the building, would probably be flooded if the rain continued.

The lieutenant who worked in the building told Abbey that the chief told the men to keep their cars on dry roads and to not worry about the parking lot. Also, he ordered them to uncover the two wooden boats that they had. The chief also told them the Network had not ordered the rescue boats soon enough.

Fortunately, the weatherman said that tomorrow and the rest of the week would be sunny days.

CHAPTER SIXTEEN

The weatherman was correct, and the next day was bright and sunny. The Atlantic had come down along with the Bay, and now the people were sweeping the water away from the fronts of the waterfront businesses.

The police had moved their cars off the dry streets and back into their parking lot. The Fire Department, fortunately, was one block higher and three blocks to the east.

Tom sat with his feet up on his desk, and Henry, who was sitting across the room, said, "You must have a crystal ball—the sales office of the boat supply corporation called and said, 'The boats will be shipped on a truck today for both the Police and the Fire Departments.'"

Tom laughed and said, "My mom called it being prepared."

Henry laughed. "She's a good judge of the weather."

Tom muttered, "And Boy Scouts."

Abbey came in with a tray of coffee and donuts. They all were happy that the rain had stopped, that the boats were on their way, and that the old boats were being reconditioned.

"The new boats will be here tomorrow," Tom said. "I'll call the Coast Guard and let them know they can arrive at any time."

When he called, the Coast Guard captain said, "My men are coming on a bus, and they will arrive in two days. They will need lodging for six at the Fire Department, and we'll need six beds at the Police Department."

Tom replied, "Will do." When he hung up, he told Abbey to alert the police and fire chiefs.

Abbey called them both directly and gave them the latest word. He would also look for a suitable place with water for their training in the new boats.

Later that day, Abbey discovered that the parking lot where they would launch into the river was still partly covered with water, but it would probably be okay for the Police and the Fire Departments to use for the latter part of their training. With the swift current the river had because of the rains, he couldn't recommend it. He would check it out, but he felt that the Carlton farm's two-acre pond would probably be more suitable in the early stages of their training.

After Henry had finished his coffee and donuts, he was back at the window with the binoculars. He announced in a loud voice, "A nice sunny day, and the two men are back at the steel bridge, hopping in and out."

Tom got up and walked over and looked out the west window. "That's the same two guys, alright. I'll have to talk to Peter and make sure that his new super steel hydroelectric scoop stands up to the explosive missiles that these guys seem to be planning."

Later, Peter told him he would check with his people, but he believed that the new steel equipment that they had installed would stand up to whatever it was faced with. It was his opinion that any missile would be swiftly swept out of the chute and back into the brook. He said, "When we tested it, the sweep was strong enough to throw out a ten-pound weight."

Tom thought that would be plenty good enough.

Henry, still holding the binoculars, said, "Come over here—they seem to have a metal container about six inches long."

Abbey said, "Our police are on the job at the hotel, but they will wait for a definite criminal attempt before they arrest anyone."

Tom said, "I hope you're right, Abbey. In the meantime, I'm going to call Pat McCormick to find out if the new EMT trucks are finished and ready to be delivered to the EMT groups in the east and west, and also whether or not the EMT people have agreed to our lease program."

Abbey said, "It's a beautiful day; I'll drive you down to McCormick's garage."

Tom replied, "Let's go. I could use some sunshine."

They went downstairs to the parking lot and climbed into Abbey's car. They both rolled down the windows and took off over what appeared to be an unoccupied steel bridge, driving past the black police sedan.

Pat McCormick's truck dealership and garage sat on the bank of the Bay, right next to the eastern row of boat houses, all six of them—including Stamford Pettibone's sixty-four-foot boathouse, which was almost at the entrance of the Bay that narrowed into the Atlantic Ocean. As they approached the garage, Abbey commented, "It has a beautiful view of the opening, and the Bay is gorgeous in the sunshine."

The dealership was a two-story building on the street with a large garage underneath for the truck repair that had installed three-foot steel posts every four or five feet along the water's edge. They were there so that no vehicle could run across the small sandy beach into the Bay. The building had two driveways; the first one went down on the west and the second one ran east downhill to the parking street level. Abbey drove down into the parking lot.

Tom said, "This sure has a beautiful view of the Bay. I wonder what this cost when he bought it."

Pat came out to greet them and thanked Tom for the two medical truck sales that should be ready to leave next week.

He reported that the EMT people on the east and west were tickled to have the Green Power Network's help. They would happily sign the leases for the repair of the two older trucks as well. They promised to visit the sales office at the Residents' Association to talk about solar panels on their new buildings, which the Network was providing.

Tom also asked about the electric transformation from his farm gas vehicles to electric. Pat answered, "It was not easy to get parts for some of the old equipment, but we've been lucky so far, and it won't be long before they're done."

Suddenly Abbey's smartphone started ringing.

He said, "Excuse me," and walked a few steps away.

The chief of police said, "The six rescue boats have arrived as well as the Fire Department's boats."

Abbey told the chief of police, "We'll be right over; we're anxious to have a good look ourselves."

CHAPTER SEVENTEEN

Abbey called Pat and Tom. "We're off to the Police Department to see the new rescue boats."

On the way over, he said, "I hope the three men the chief has assigned to each boat will take their training seriously, especially considering all the expenses that we have spent to make this possible. When the Coast Guard trainers get here, we have to show them the highest hospitality."

Abbey parked next to the chief's car. The chief came over as excited as a boy on Christmas morning and said, "They threw in a freebie for the sales. Look, it's a three-story, steel mobile rack for six rescue boats, and motors so our cruiser can pull it. Can you believe they gave us a freebie?"

Tom and Abbey shook his hand. Tom said, "I'm glad you like it."

The chief said, "A fine trailer like that is expensive. They are a generous public service corporation."

Abbey said, "I think I found a place where your men and the Coast Guard people can practice."

The chief said, "Good. We'll devote every spare minute to perfecting our skills in these boats. In a serious rainstorm, like we had the last couple of days, we're going to have to be good at what we do."

Tom said, "That's exactly why our Green Power Network bought them for you."

Two days later, Abbey and Phoebe were at lunch at the snack bar. Abbey got a call from the Police Lieutenant Brian who said, "I'm on the second floor in the back bedroom facing west, get here as quickly as you can. I think it's going down."

Abbey jumped up and said, "I've got to go—keep up if you can."

He took off. Abbey was soon just behind the lieutenant who handed him an extra-large pair of Navy binoculars.

He then said to Abbey, "Look at these three points. First, across the brook to the deep woods. Then the top of the tree, to the right of Hill Street beyond the bridge, on top of a telephone pole, and there—it's the steel bridge itself."

Abbey took the heavy binoculars and scoured just as the officer suggested. He viewed the biggest tree in the woods and saw a high-powered camera and an officer at the very top of a telephone pole in a tent on Hill Street with another mounted camera. He then turned to the steel bridge, and it seemed they were right next to them—they were that close.

The lieutenant said, "The two high-powered cameras can see the building real clear, including the brook at the scoop opening, and we can see every move they make."

The lieutenant continued, "We couldn't get sound without giving away our positions, but we can see them close enough to judge what they are saying. Their actions show me they are loading two yellow projectiles just under a foot long. They chose two because one might miss and the other will go through the scoop hole."

Abbey asked, "What makes you think the deadline is imminent?"

The lieutenant answered, "They're both nervous and fidgety. I can see lips moving and tension on their faces with these high-powered binoculars."

Abbey said, "I'll get Henry and Tom on the phone immediately. Don't let what I'm doing hold you back; you go for it if necessary."

Abbey called Henry's office and said, "Put Tom on the phone—don't talk, just listen."

The lieutenant said, "They have launched two missiles in the water. One is in the hole of the scoop, and the second is in the scoop. The man in the tree with the heavy-duty camera says the first explosion is in midstream while the second missile explosion is stuck in the opposite bank—¬no harm that I can see to the building."

The officer on the top of the telephone pole said, "Four officers are racing to the steel bridge. One man apprehended the second man and couldn't be seen. Two men were running down the street looking for him. They only encountered a disabled man with a dark coat, a long gray beard, and grey hair. They held him and quickly searched him, found nothing, and released him. It was a successful defense of the building and the property, with the apprehension of one culprit."

Abbey said, "Did you guys hear that?"

Henry answered, "Tom and I heard it loud and clear. Good we got it all on film, too. Congratulate the lieutenant for us."

Abbey said to the lieutenant, "I love your surveillance gear."

He responded, "It's all on loan from the Navy."

While the lieutenant was putting away his gear, Phoebe came up into the room, standing quietly and listening to it all. The lieutenant was looking around and asked her, "When are they going to give you furniture in your classrooms?"

Phoebe put up her hands and said, "I'm told it's all in order of priority."

The lieutenant looked at Abbey, who said, "It's all a matter of priority, my friend." He turned to Phoebe and said, "I'll talk to someone, I promise."

Phoebe answered, "I've hired a young woman from the neighborhood;

she's an experienced receptionist before she got pregnant and had to raise two children—a boy and a girl. Now they're just out of second and third grade, so she's going back to work. She likes the idea of being able to walk to work here in the neighborhood and then pick up her kids after school.

"We'll carry on in the first room on the left downstairs. It will be a reception room, and we'll have orientation meetings for new employees and students there.

"Henry has agreed to provide the right kind of instructors from his staff for each of the classes. When we get some furniture, then I can advertise for students.

"But I still must sit down with Tom and decide on the cost of the various subjects. They will be cheaper than trade school or a college, I hope. But I have a question that relates to what we will be doing.

"While I was talking to Tom, I noticed a large carton of small green bottles, and I asked him what they were for."

Abbey said, "We're going to install them on all our electric charging stands instead of the white light that's there now. Then, police, firemen and other rescue workers on rescue missions will know where to get a free electric charge without having to return to the station. It gives them more time to rescue people in trouble. There will be no charge to their departments."

Phoebe said, after some thought, "With all of that on his mind, I can understand why my school teaching assignment isn't a very high priority. I'm not upset about it. Let him know that, Abbey."

Abbey responded, "My first class will be on the creation of electric power. Getting started is a matter of priority, and I can't wait."

The arrival of the Coast Guard—twelve people, all in uniform—got the attention of the police station. The police chief had noted on the bulletin board the eighteen men chosen for rescue boat duty.

The police showed the guards their quarters and gave them a tour of the station. The guards and the officers ate together and set up a training schedule. The first two days would be in a classroom with one boat and a barrel of water with an electric power outboard motor sitting in it. With the rescue boat and its on-board equipment, knowing the batteries and keeping a strict routine were the first things they would learn to depend upon.

Next, it would entail turning the motor on in the barrel, reading the two clocks on the hood—one for the time of day, and one for the power remaining in the ninety-minute capacity before recharging would be necessary. Three men were assigned to each of the numbered boats: One on the outboard motor, and two to handle the boat and control the rescued persons.

The two clocks on the outboard motor's hood showed both the time of day and timeline for the rescue operation, which indicated the remaining operating time before it needed an electric recharge on the motor. The other equipment was explained and demonstrated by the guards.

Abbey came to one of the meetings and explained the choice of practice areas. First, they would have a tour of the forty-five-acre Agrivoltaic solar installation and then go next door to the pond for their preliminary experience with the rescue boats.

When the guards thought the police rescue teams were ready, they would move to the river launching site for more difficult rescue operations. The basic loading and unloading of the six boats and their outboard motors would first be done in the police station's parking lot.

The guards worked the officers from dawn until dark, only resting for their box lunches from the farmstand that were delivered by Nancy.

They were in the river loading area by day three, and it served as their second learning experience. It turned out to be hard fighting the current and the fast-moving, slightly-flooded swift river.

They were powering upstream and familiarizing themselves with the low-lying homes and businesses in the floodplain. Then they traveled down to the Bay and fought their way back upstream.

They also found four green recharging stations as a backup along the way.

The police chief observed the last several days of their training from one of the old wooden boats they had, and he made several contributions. They practiced loading and unloading rescued people of all ages and got a feeling for how difficult and time consuming it was.

The next day, back in the classroom, the chief gave each man a Coast Guard's patch for their blue uniforms. The patches were provided by the guards.

The Cliffside mayor, who attended the meeting, gave the ranking Coast Guards a check from the Green Power Network for $25,000.

The Coast Guard then moved on to the Fire Department for a repeated performance of the rescue boat training at the preliminary farm location, and then back to the river launching area.

The mayor also repeated his speech of appreciation for the guards' public service and presented on behalf of the Green Power Network another check to the Coast Guard from the fire department for $25,000. The fire crews also received patches like the police did.

Later on, the mayor asked Tom, "How can the Green Power Network afford such generosity?"

Tom answered, "Our forty-five acres of Agrivoltaic solar panels produce six megawatts of power an hour. We sell what we don't use at the farm to the Cliffside grid at the going price-per-hour, which increases during an emergency. Our operating expenses come from the power provided by the eight-tower wind farm fifty miles off the coast of the Atlantic, which is part of our Network.

"Our mission is to return the proceeds, after costs, to the benefit of the city of Cliffside and the safety of our citizens in case of a serious climate emergency."

The mayor said, "That's a wonderful public service. I'm glad to be a part of it. You can call me anytime for any help that you may need."

A few days passed, and it rained during most of them. Abbey found himself on duty in the ground level room behind the snack bar. The red light over the incoming intelligence microphone lit up.

Kirk Forest, a familiar voice, said, "Joe, I came by to tell you we cancelled our helicopter straw attack over the farm. There's just been too much rain, and nothing will burn as we had hoped. Also, as you know, missiles from the bridge were ejected by their automatic scoop. I got away alright, but your man Ciccone was arrested. Our lawyers are taking up his case and will arrange for his release—don't worry about him, they'll take care of it. We have a backup plan that's even more destructive because it will come during a natural disaster."

Joe Hendricks said, "How can you guys predict a natural disaster?"

Kirk answered, "We have national connections, and we know when to move and when not to move. We put a couple of our science people on the problem, and they have developed a plan. It's going to take some dark money to do the part; I'll get back to you as soon as we have more details."

Joe said to himself, "I wonder what could be bigger than a straw burning on the farm. I'd better put out the word on this one."

Abbey discussed it with Henry and Tom. They agreed it sounded ominous. Tom said, "I'll call Pat McCormack and make sure the new electric trucks have arrived, been serviced, and put into EMTs hands, and that they have been trained to use the new equipment. Also, the old trucks are still being redone.

"We've gotten good news from Morning Sun. They have solved the problem of the self-reliant silo service for homeowners who want emergency protection. They're going to send the sample up to the Association building and sales office. George called to confirm the breakthrough and said, 'We'll set a bright green plastic drum three-feet high and twenty-four inches across with the Green Power Network's name on it and advertise seven-day power protection.' They also reported that Morning Sun is now building the two garages and they will have solar panels to add to the new seven-day protection. He said he will call Stanley and alert him to the fact that the new stuff is coming in."

George also mentioned that he wanted to talk to Tom and Henry privately sometime at the farm for a new addition to the array.

Later, Tom called George and asked the reason for the top-level meeting.

George answered, "A new idea—a new addition to our vegetation and a way to increase the efficiency for our Agrivoltaic setup."

When Tom and Henry arrived, George said, "Let's take a walk in our field of products—vegetation and solar."

Henry said, "Lead on."

George started walking among the Agrivoltaic solar panels and the growing plants.

He said, "Let me paraphrase, the United States hopes to decarbonize cities by 2034, according to the Biden administration. Under Trump, who knows. But to do so, they must feed a nation on a million acres of Agrivoltaic. That's because plants rely heavily on solar energy, which should reach forty percent of our green solar power, and it will be lighting our homes and feeding us by then. They have created and funded six research projects to see how to do it. One idea is a growing project of vertical solar panels that allow cows to graze between them.

"A new consideration for the Agrivoltaic farmer is to have equipment that tends the crops more efficiently and is labor saving. It seems they have already developed robots who can pick out weeds and not harm the crops.

"The robot uses three high resolution cameras to peer down. It's assisted by strobe lights and a computer, which creates a digital image of each seedling. The computer arranges for the robot to put a dot on the isolated weeds and withdraw it. The whole process is instantaneous and happens in a fraction of a second.

"The farm robot rolls down a field at less than a mile an hour. The owner, Todd Rinkenberger, has made the robot in Chalara, California.

"I'm recommending we get one of their robots and put it to the test right here on our installation."

Henry asked, "Has anyone thought to ask how much this robot costs?"

George answered, "Stewart industrial technology in Salinas, California, will tell you if you call them. Their robot can weed an acre in an hour. In addition, they'll help you find tech people who will apply their knowledge to agriculture. You can also call the western growing association in Salinas for more information, and it would be nice if you could get some pictures to look at."

Tom said, "I'll call them right away and get the particulars. We may even become their eastern distributor."

It was nine in the morning, three days later, when Tom and Henry were in their second-floor offices, each contemplating the day to come.

Tom said aloud, "We're really indebted to them."

Henry looked over and asked, "Besides the Forever Blue Public Trust, who are we indebted to?"

Tom observed, "They saved our backs at the City Council."

Henry responded, "Oh, the dairy people—well, I guess you're right. We do owe them something."

Tom said, "We're having the Morning Sun contractors put in a large set up of solar panels on the Cliffside dairy cow barn. It is four-hundred feet long and they have given them a twenty-five percent discount on covering all electric needs around milking and cooling. He's also converting his trucks to electric power. Morning Sun also assures him that the grid will send him a nice semi-annual check with the leftover power that they were able to receive from him. But that's only true for that one dairyman. How about the others out in the rural areas and on the Reservation? Let's work closely and make their prices jump up and down to make a steady income."

Henry asked, "Who's going to do this investigating?"

Tom thought and then said, "Phoebe Sinclair, along with my mom."

Henry laughed. "Two female presidents would certainly have the easiest time when deciding what we're going to do."

Tom answered, "Who's more interested in milk than a mother? I'll ask Phoebe myself."

He picked up the phone and called Phoebe, who was at the Painted Lady, and explained what it was he wanted to do and asked if she would take the job.

Phoebe thought for a minute and then said, "Yes, I would be happy to do the job, and I will ask your mother if she's willing to help."

CHAPTER EIGHTEEN

Tom looked at Henry, smiling, and said, "Let's print up some business cards for the ladies."

Henry added, "Good idea, and let's have the cards embossed with Green Power Network, incorporated. Each one will also have to have their names and positions in the company, along with the plant's address."

Tom said, "You know, I heard many years ago that in the northeastern states, among the greatest number of domestic animals kept by farmers were not the cattle but sheep. That being true, I wonder why it's not still the case."

Henry answered, "The modern man and woman prefer beef and pasteurized milk. You can also produce more sellable products with beef than with sheep, especially with the country's population growth."

Tom countered, "But today's dairy market seems to fluctuate a great deal, despite a large quantity of beef and milk products available. I hear many farmers are quitting the family farm. The average farmer today has six-hundred acres, eighty cows, and is considering leaving a business that has supported his wife and family, and in many cases, kids who might want to inherit the family business.

"The family farm usually bottles their own milk and makes ice cream and chocolate milk. The cows seem healthy, and the pasture is plentiful. Others, who have left farming, say there isn't enough money in it for the small farmer anymore. Their children feel that they can make more money in other businesses and work nine-to-five during the week with holidays off.

"The farmers who find they only have fifty cows are behind the times and would like to get into the new technologies and equipment, but they don't have the time or the money to invest in the new demands for their dairy farms.

"Also, the housing shortage has enticed most farmers to give up their land for housing development and the money it provides. Realtors are offering big money these days for pastured lands. Other types of businesses are putting the same property to more profitable uses, such as farms for growing beverage developers."

Henry said, "I heard a statement the other day. 'Get Big or Get Out' is what's describing the new future of the dairy industry. The persistent oversupply of milk in the U.S. depresses prices for the dairy farms. The hundred-cow farms are too small to operate with the same offerings and enticements offered by the thousand-cow farms. That's where they get the phrase, 'Get Bigger or Get Out.'

"For instance, in 1950, the state of Vermont was home to eleven-thousand dairy farms. Today, there are only five hundred and eight. In many cases, those small farmers are paid less money and their costs are not covered. In addition, their products are exported around the country and the world.

"We must consider technology, and especially marketing technology and how it has always been the keys to American success. We are known for it."

Tom reiterated, "The new tech solution is, in my opinion, green power. It is the most profitable investment that a dairy farm or a farmer can make today. They can use their own green power to take care of the needs on their farm, their home, and their car, and then pass the surplus along to the grid for an additional income. In an emergency, you can get an increasing amount of money for your surplus power, too, all of which means an improving financial situation for the local farm and dairymen.

"I've explained all of this to our two lady salespeople, and I suggested they visit our new sales office in the association building and talk to Stanley, our new sales manager, who I'm sure will have more tips for them."

Henry said, "Let me change the subject. I'm still curious about what we're going to do about the robot for George?"

Tom answered, "I have called them, and the girl answering the phone promised to send me some literature and pricing."

Abbey came in just then and said with some relief, "Our villain, Kirk Forest, has been quiet lately, but our police intelligence are watching and listening. They are searching for clues in all directions."

Changing the subject again, Tom added, "I forgot to tell both of you: George and Nancy have become engaged. No date for any wedding has yet been set, though."

Henry laughed, saying, "I wondered when they were going to take the leap. My wife thinks the fall is preferable."

A couple of weeks went by with business as usual, when Stanley came into Tom's office with the month's sales record. He was all smiles.

He said, "We have five new solar panel home sales. Two of them included two-thousand dollars each for seven-year emergency coverage, with our new green silo protections. Three of those sales are new family dairy sales of fifty cows or less. They love the idea of getting a check from the grid at Christmas time. It was your mom who told them about her leftover power checks at Christmas time. That was encouraging news for our new sales."

Stanley continued, "The ladies are doing okay; your mom is learning quickly. Their sales are adding to our sales at the showroom. Green power will be well on its way in this area."

Henry added, "Our home plant energy and sand silo volume has picked up in sales, too. The high-energy sand silo is becoming popular storage for many businessmen. If our seven-year home, self-reliant, bright green silos take off, we will be the biggest non-lithium storage system in the northeast. George told me that Morning Sun filed a request for a joint federal patent and has added Green Power Network as a co-owner. I thought that was a nice gesture."

Stanley added, "The happy ladies on the road keep producing week after week."

Tom added, "Stanley, curb your enthusiasm and tell the ladies to come back tomorrow; we're going to have a meeting downstairs at lunch together."

Stanley then added, "Just as well—Hurricane Jenny is now off the Florida coast and may come closer and give us more rain this week. We better get the sandbags ready. The brook may start roaring like the last time we had a hurricane that came up to the northeast."

Henry asked, "What sandbags?"

Stanley answered, "The ones in the guard shack. We block Hill Street steel bridge when the brook begins to get over the top and threaten the coal factory. The brick wall on our side of the road is three feet higher, and with the sandbags we connected, the wall on the factory side is three feet higher. So, the brook runs right past, even though it's overflowing the steel bridge. We did this the last time we had a hurricane and the brook threatened to go down Hill Street."

Henry continued to ask, "Why doesn't the water continue to wash down Hill Street?

Stanley answered. "Charles Morgan, the man directly across the street from us, puts sandbags between him and his neighbor's walls as we have done on this side. So the brook continues along its merry way."

Tom asked, "When did you guys do this last?"

Stanley continued, "Since your dad, Tom, rebuilt his pond dam, the brook has always run higher in a heavy rain. We built the walls up on both sides with the rocks that came out of the old dam that your dad replaced with concrete.

"The last big hurricane was ten or twelve years ago. The sandbags worked fine. The city thought it was a great idea, and it saved a lot of people down the street from being flooded out. the coal company management liked it, too."

Tom said to Henry, "Tell your men about the sandbags in the guarded shack. Make sure they're in good condition and buy more if you need them. Be prepared to build a two-bag thick sand bag wall across the way as Stanley describes it.

"Stanley, you go across the street and talk to your friend and make sure he's prepared to do just what we're going to do."

Stanley said, "I'll do it as soon as he comes home from work. He's a good guy; you'll like him."

Tom commented, "The weather report says we may get more than one day of heavy rain if the hurricane continues to move closer to the coast and further north. We want to be prepared. Abbey, about your friends at the police station and the firehouse—can you call them and make sure they've got their rescue boats at the ready and check to see if they've heard any more from Kirk Forest?"

Henry added, "There's a guy in the Association who has a pickup truck with a plow on it. I'll ask him to put the blade behind the sandbags to firm them up. We'll buy more so we can alternate the bags and double them. We'll also make a detour sign to put up at the factory street corner."

Tom said, "That should protect our building. I'll call George and bring him up to date. Abbey, tell the Red Cross they may use the three rooms downstairs for injured persons if they wish and offer the Salvation Army room on our property to erect the tents for cots to help flood victims rest. Okay?"

Tom didn't hesitate; he wanted to be prepared if the hurricane hit. He was going to go back to his desk and prepare an agenda for tomorrow after he talked to George.

But first he said, "Oh, by the way, Stanley and Henry, make sure you congratulate the two ladies when they come back."

The next morning, Tom spoke to Joan, who was the lunch supervisor, about his plans and then talked briefly to Abbey in the backroom. He mentioned that the Coast Guard sent their thank yous for the fine conditions.

The microphone was picking up in the back room. Joe Hendricks had gotten in touch with his fuel supplier to make sure he would have a delivery during the storm, and he was assured that his deliveries would be on time. Satisfied with that, Joe was happy to hear from Kirk Forest and they went back to work. The mic clicked off.

Tom and Abbey went back out front to help Joan set up for the lunch meeting. But both wondered what Kirk Forest had just completed.

At eleven-thirty, the staff started to arrive. Tom and Abbey sat down at her bar table and quietly drank a cup of coffee. They recognized and welcomed the people as they arrived and suggested they take a seat. They would have the meeting before lunch.

On the dot of twelve, Tom tapped his coffee mug and called the meeting to order. He thanked them all for coming on such short notice and on a rainy day and said, "We'll begin with the department heads' reports."

He introduced George Ralston, chief engineer on the solar array at the farm. George began by talking about the new home, their bright green plastic silo, its insulated container that was three-feet high by twenty-four inches in diameter that would go with the generating equipment when a person or family bought self-reliant, seven-day emergency protection.

"It is the first of its kind by Morning Sun Incorporated and is now for sale at our associated building sales office," George said.

The audience clapped enthusiastically. George told them they would be able to see it in action later.

Tom thanked him and introduced Henry. Henry reported on providing two new EMT electric trucks free of charge that were fully equipped, as well as the two repaired and converted used electric trucks. He also mentioned that they were building two truck garages, one for the East County EMT, and one for the West County EMT. Each would have charging stations and solar power units in their garages, using the new bright green silo self-reliant power.

In addition, they had provided, without charge, six blue police rubber rescue boats, fully equipped with electric outboard motors, and six red fire rescue boats for the fire company. The company also paid for the Coast Guard training for their personnel. The reaction was overwhelming but someone asked, "How can we afford all of that?"

Henry responded, "By getting and selling Agrivoltaic to our farmers, thanks to our own forty-five acres of green solar power. We'll use what we need, and we'll sell the rest to the grid and use their money for our protection during an emergency in Cliffside."

Henry also reported, "Solar panel sales to new dairy companies that will use the power to milk their cows and run their business and send the surplus power to the grid for sale. This, of course, will improve their financial situations."

He also took a moment to congratulate the traveling sales ladies, Priscilla Carlton and Phoebe Sinclair, for their traveling sales efforts.

Just as Tom rose to introduce the next speaker, Stamford Pettibone entered the room. Though surprised by his presence, Tom introduced the owner and director who owned eight wind towers in the North Atlantic.

Everyone applauded, he bowed and thanked them politely, and sat down next to Mrs. Carlton.

Tom asked, "What brings you to us on such a rainy day?"

Stamford answered, "I came with my yacht captain and two crewmen to take my yacht to sea. I'd rather face a rough sea with that yacht than have the yacht banged around in my boathouse when it's struck by an Atlantic storm surge."

Henry asked, "Do you expect that rough of a sea?"

Samford replied, "The national oceanographic experts have been keeping an eye on Jenny, the hurricane in the Atlantic. Right now, it is just a tropical depression off the African coast, but it may come closer to the U.S. coastline. Their clue is the heat of the ocean waters. They are as high as eighty degrees. That can have a fast-changing impact on the hurricane with winds as high as a hundred-and-thirty miles per hour. A lot depends on how strong and expansive the Bermuda or the Azores patterns get. They act as a steering wheel for storm waves and hurricanes, even if the hurricane is off the coast and not inland. There could be dangerous impacts, like storm wave surges from seven to eleven feet high."

Stamford continued, "The National Weather Service suggests: don't mess around with this one.

"Having a strong surge like that will be pushing it into our oncoming swollen river. It could raise the water level considerably in the Bay, and therefore, damage those businesses, homes, and boat houses that are directly on the Bay. Combine that with the hurricane rain and the winds that you've had for the past couple of days, the hurricane could be a calamity.

"The experts know a traveling hurricane thrives on warm water. It fuels water vapor that makes it stronger. This season, the seas are warmer—it's twenty-five degrees above normal, according to the averages between 1992 to 2020. Basically, six degrees of warming is a seven percent increase expected for hurricane rainfall."

Abbey whispered to Tom, "I'll bet that's what Kirk Forest referred to when he was talking to Joe Hendricks about having very powerful friends."

Stamford continued, as he accepted a snack and a drink from Joan, "We'll set out to sea as soon as the crew says the yacht is ready for a rough trip."

Tom asked, "Can you call me if you have to interrupt your daily power shipment to us?"

Stamford answered, "Yes, of course, but I may need to use some of it if the farm has any kind of damage."

Tom said, "If you need it, tell me how much, and I'll put it in the sand energy silo so you can draw on it when you need it."

Stamford replied, "Good."

He finished his burger, fries, and coffee, then moved from his chair, waving goodbye as he left.

Tom said to the fascinated group, "Enjoy your lunch, and take your time. I'd like to ask the staff leaders—Henry, Stanley, George, Peter, Abbey, Priscilla, and Phoebe—to stay after they've eaten."

When they were alone, Tom said, "Apparently, Stamford has access to expert oceanographers and advanced information revolving around the weather at sea, and the possibilities of this hurricane moving to the coastal U.S. area. Bermuda and the Azores are the steering wheels for storms—real scary at the least, but it gives us some advance warning. Be sure your departments are storm proof for the next several days. Also, be prepared to help any victims of the rain, wind, or storm surges if there's flooding around the Bay. Stanley and the ladies, please no sales effort for a couple of days. Let's get this storm behind us. If the people south of us on Hill Street are flooded in, how will they get over the swollen roaring brook to the Salvation Army tent shelters?"

Stanley replied, "The Association has had this problem before and has developed a system. They use two extension ladders lying down, side by side, to cross from our wall to the opposite wall on the Southside, with wooden planks to walk on. Then, they stretch a tight, heavy rope from a tree to a post on either side and draw it tight so people have something to walk with and lean on as they come across the ladders and planks. I'll make sure it's available."

Tom said, "Good, give your Association members a gold badge for their uniforms for being prepared for any event. Thank you."

Priscilla interjected, "Sounds like you guys got it under control. I'm anxious to get back to my farmstand. We may need to provide some food and beverages or something for hungry people, and I think I can be of service there."

She turned to Phoebe and asked, "How do you feel about that?"

Phoebe looked over at Abbey, and he said, "Phoebe and I will work in the tents and help the people coming to our rescue tents. They will probably be asking for fresh vegetables or dairy products."

Tom said, "Great," and turned to George. "George, keep us aware of the overflow on the pond next to the array. Henry will hopefully get our sandbags in place before that dam overflows. There are a couple of small streams on the mountain that feed that pond, and they could flood quickly if this rain continues heavily.

"Also, George, give us all the power you can find for the next few days—I want to fill our energy silo. Our rescue boats and EMT trucks will be needing recharging every ninety minutes when they're operating. You, Stanley, and Peter, will be with me in the plant, probably for twenty-four hours once this whole thing gets started."

Tom called in Joan, saying, "Your husband is a caterer, correct? Ask him if he can get Abbey and Phoebe and their staff food and equipment in the tents if he can, okay? Our company will pay for all his expenses."

Joan replied, "I'll call him right away. I'm sure he'll be happy to help in any way he can."

Abbey opened his cell phone and called his friend in the circus tent business to be sure that they were prepared and on their way. They announced that they were using footers on the fencing on the factory side to support their poles on the east side of the building.

Tom looked around to all of them and said, "This is probably the first major challenge for our new emergency service by the Green Power Network. We created it to show that the Green Power Network can serve our community well in an emergency. Let's do it."

CHAPTER NINETEEN

The rain was just as heavy the next four days. The tents were up and secured with two steel fence posts, all the sides flapping down, and the wooden posts secured. The flaps were held down and secured by long steel pipes driven deep into the ground.

The Carlton's staff were moving into the larger first tent with food tables and chairs, as well as kitchen utensils.

The Salvation Army people were moving into the second tent with cots and blankets. The big white Red Cross truck pulled in next to the entryway. Workers opened the rear doors and began to unload materials for their infirmary.

Phoebe was showing the two nurses the three lower rooms in the entryway to use for their services. They agreed that the rooms were good spaces for injured people on cots.

Henry's staff, along with Stanley, were loading sandbags onto wheelbarrows and carrying them over to the Hill Street roadside bridge.

Abbey and Phoebe were working steadily with the caterers, setting up their kitchen, the tables, chairs, and the big tent.

Tom was in his office watching the National Weather forecast on

TV. The hurricane was moving north but still wasn't close to the New England coast.

Tom gave special instructions to his secretary to make four signs for the locked doors to the working rooms with electric power. The signs would say: "Danger—Do not enter. We work with raw electric power." He asked her to make them in red paint.

George called and reported the pond was still rising but had not reached overflowing yet. He said he would call back again at noon, at five, and again at eleven p.m. to report on the dam's condition.

The weather forecaster was now beginning to discuss the Atlantic Ocean's heat system that spring. The expanding winds over the North Atlantic water currents were known as the Bermuda and Azores highs.

High pressure contracted the strength of the trade winds. The weaker winds that blew decreased the evaporation at the surface, which allowed the temperature of the water to increase rapidly. Roughly forty percent of the world's oceans were experiencing marine heatwaves.

The increased ocean warmth could help usher in the hottest year on record beginning in July. The warmest July on record in the continental U.S. was twenty-one degrees above the average recorded low.

August was now here, and they were just coming into the hurricane season. According to a Colorado State University team, annual upgrades and strengthening was attributed to extreme warming in Atlantic Ocean waters. Greenhouse gas emissions were attributed to the evaporation, which was what was heating the ocean water.

Tom was surprised when the weatherman mentioned a new set of initials to remember—AMOC. The weatherman told them that warm salty water moved north in the Atlantic from the tropics along the Gulf Stream on the east coast of the United States. It then cooled in the Canadian provinces and sunk and returned south.

The faster the water moved, the faster the water turned from warm to cold. The greenhouse gases caused more global warming, which sped up the melting of the Greenland Sea, putting fresh, warm water into the mix, which slowed the cooling and sinking process. So, the oceans stayed warm longer in the North Atlantic. All this caused serious warming and climate changes in the Canadian provinces, northern New England, as well as—believe it or not—northern England and Europe.

This continuing climate uncertainty caused a recent special British publication in The British Journal on nature communication.

Tom said aloud, "Well, I'LL BE DARNED!"

He looked up and said, "Would you have believed? ALL THE WAY to Europe."

Henry thought a moment then said, "Tom, I spent fifteen years as a lineman on the telephone poles of this city restoring power to families because of a National Weather emergency. If I can do anything now, like fighting the carbon isolation of our atmosphere, to prevent national emergencies and improve our weather, I'm going to learn what I can do and do all I can—like being here, at the Green Power Network. I'm going to give it my best effort."

Tom answered quietly, "You're right, but I had no idea that it was all consuming with the countries around the Atlantic. I didn't believe there was such an immediate connection with other parts of the world."

He turned off the TV and said, "I'm going to go around and see how our preparations are coming." He left the room.

Henry left behind him and made his way down to his staff to check on the silo sales during the storm.

Tom was in the tent checking with Abbey and Phoebe when his cell phone rang. He picked it up to hear George shouting.

"I'm at the pond. The dam is overflowing—like killer waves. It's really coming to you. I hope your sandbags are in place. The wind is coming on strong so the hurricane must have come closer to the coast."

Tom called Henry on the phone immediately and said, "Come over—the water is coming over the dam in heavy waves. Check on your sandbag people at the Hill Street bridge."

Tom called Stanley next and shouted at the phone, "The dam is overflowing with heavy waves. I hope you're prepared."

Stanley responded, "You're telling me; I'm looking at it, fly boy. I hope the guys at the bridge have their sandbags in place. I heard the guys on the other side of the bridge are still assembling and placing their sandbags and getting soaked while doing it—but they're being successful in keeping the water from flowing down Hill Street."

Stanley then said to Tom, "I'm a little concerned with the gully where the brook makes its turn east towards us. The west side of that gully is low, and I'm afraid a percentage of the water is flowing straight down that field towards those homes on the west side of Hill street. I would suggest you call Abbey and tell him to alert the police to the problem—maybe they can help those people from being flooded somehow."

Henry's people reported that their side of the steel bridge was dry and that the sandbags were effective. The street water that was flowing down part of Hill Street was being contained by the sewers, which were piped under the bridge into the brook—so there were no problems. It was still raining hard, but the brook was well contained by the sandbags, even though it was fast and heavy and more than a foot above the bridge surface. They had another eighteen inches to go before the water was over the sandbags and they were flooded.

As requested, Abbey called the chief of police and made him aware of the deep water coming from the bend of the roaring brook. The chief said he'd do what he could.

Abbey then told the caterer to be prepared with hot coffee and food, because the pond dam was over the top and the people would be arriving shortly from the west side of the brook from over the extension ladder bridge.

The police reported to Abbey that the heavy rain had made several streets impassable. The road surfaces city-wide would be flooded, with all sewers beyond capacity. Working people would find flooded roads blocking their normal driving routes home. The police and firemen warned people not to drive through flooded streets using their loudspeakers on their rescue boats. "Leave your car on high ground; we'll pick you up and get you home or to a safe place by electric rescue boat."

Even small, usually ignored, half-dry streams were blocking roads and threatening first floors and cellars. The interstate in and out of Cliffside was blocked by water on the downhill areas. People were finding their homes under water up to the first floors.

The six police rescue boats and the six fire rescue boats were communicating and cooperating to get people out of danger and down to the City Hall service areas. Other municipal buildings were being put into service, as well as Red Cross areas that were helping city center flood victims.

The Fire Department had all six boats out, and some of them were bringing students from the college back to the firehouse. All of their trucks were out and on high ground, and the firehouse was acting as a shelter.

Cliffside had primarily hilly streets. One would think it would be an advantage, but once the sewers were filled and they couldn't empty enter the Bay, they backed up and the streets around the sewers turned into rising ponds.

All these conditions continued well into the night with increasing numbers of people coming into the Green Network tents for warm meals and resting places for their children.

Abbey and Phoebe took a break at two a.m. and occupied two of the Salvation Army cots and some blankets. They slept until six-thirty a.m. and were awakened to help feed breakfast to the children and then to the parents. Four uniformed policemen kept order through the night. Two slept in their police cruisers on four-hour shifts. Henry and Peter's staff in the building were up all night helping the Red Cross nurses take care of the injured people who had found their way to them.

Tom was still in his office checking the weather.

The river had risen to a depth of twenty-one feet in northeast New England and had received between five and seven inches of rain in the last forty-eight hours.

Henry's staff in the plant had been making hydro-heated power from the silo and adding the wind farm power during the rain storm. They had been keeping the green charging stands well supplied and serving the rescue boats. The electric trucks with the EMTs were working effectively in the east and west counties, though they needed to recharge every three-and-a-half hours.

The grid announced that it had 20,000 customers without power, and their teams were busy restoring power as quickly as they could.

Henry told Tom, "Here it is, the third morning, and the rain is still as heavy as it's been. George's Agrivoltaic array is functioning just as well as expected, our energy silo is working fine, and Stamford's wind farm is supplying us with nightly power without interruption. All this is the product of our green power, which is totally carbon free."

Tom added, "The dam has overflowed, the river is well over its banks at twenty-one feet, and the Bay is quickly getting deeper because of the constant river increase. We're doing fine right now; let's hope it continues.

"The weatherman is announcing that the sun will come out this afternoon. That means our Agrivoltaic array will start producing power again in about an hour. I'll have to confirm it with George."

Henry replied, "Hold on, my super entrepreneur—there are a hell of a lot of cleanups that will have to be done after this rainstorm. I'll bet that Pat McCormick's workshop is full of water, maybe even with live fish in it right now. The water leveled businesses at the edge of the Bay, and people have lost hundreds, or maybe thousands, of dollars in merchandise."

Tom admitted, "That's true. Our bayside shops and dark boats, as well as the boat houses, have probably experienced a lot of damage. But we have earned an honored place in the town with our sandbags on the steel bridge."

Henry answered, "Hurray for us, for we changed it, worked hard, and made it happen. But don't forget to send a nice letter and a check to our caterer, the Red Cross people, and the Salvation Army staff. We've had a lot of help from good, hard-working people. Also consider our hard-working employees."

Tom said, "As usual, you're right, Henry, and I won't forget any of them."

Right on schedule, the sun shone bright, and it became beautiful early in the afternoon—just as it was predicted.

Abbey called the chief of police and the fire chief, as well as two EMT medical truck units. He congratulated them on having done a fine job but requested they keep the boats in the water and the trucks continuing to hit the pavement.

He told them, "There is no charge for the power they use or any equipment they may need. If anyone needs service on the electric outboard motors, they can take it to Pat McCormick's truck garage. He can place any charges on the Green Power Network account.

"When your job is finished, please send us a report of your services and the costs of each area. Send it to Tom Carlton, the director of the

Green Power Network. Take pride in the fact that your services have added no carbon to our atmosphere during this whole experience."

On Abbey and Phoebe's suggestion, Tom asked Henry, George, Peter, Stanley, and Lieutenant Gibson to breakfast.

Tom asked Edward and Joan, the caterers, to put on a fine breakfast with hash-browns, sausages, eggs over easy, orange juice, toast, coffee, and danishes.

Tom opened the breakfast by thanking the leadership and their staff during the storm. He told them they would all receive a bonus as a congratulations.

He continued, "I want to announce two additions to our product line as well. First, our new self-reliant, green, seven-day emergency kit will be added to our home solar panel units for an additional cost of nineteen-hundred dollars. The second thing is we're going to add seventeen more acres to our Agrivoltaic array, starting immediately. We suggest giving our staff a nice raise; they've earned it."

Abbey led the applause. But he was pulled away by Joan to the secret microphone recorder. Abbey got up, said nothing, and left. Tom noticed and nudged Henry but said nothing.

Abbey rushed to the backroom of the snack bar and turned on the recorder. Kirk Forest's voice was telling Joe Henricks, "The Green Power Network Truck will be delivered tonight at midnight. There's a tank of corrosive metal with a pump and two nozzles in the back. Make sure your driver wears his green uniform and works late at night. First, pump it into the green fuel supply at the police station, then at the firehouse. He must finish by dawn with the local neighborhood's charging stations, and he must leave the truck and his uniform in the river."

Joe asked, "Where the hell did you get a Green Network delivery truck?"

Kirk laughed and said, "We bought the same model truck as it and painted it to look like theirs. I'll say goodbye. You won't see me again."

Abbey sat back and said to himself, "Liquid corrosive metals injected into the pumps and chargers could put us out of business for... I don't know how long."

He called the police chief immediately.

Back at breakfast, George got up and said, "Thank you to everyone. It was a pleasure working here with you all. I guess you all know by now that I'll need the extra money. Tom tells me Nancy has very expensive taste."

They all laughed.

Then Lieutenant Gibson walked by the head of the table and said, "I'd like to introduce Robert Thompson, United States Marshall, who will take care of Kirk Forest as soon as he's arrested. He's the man who tried to bomb your plant a couple of weeks back. Well, we're not going to keep him. Our Federal Marshall tells me he has much larger federal charges facing him in Louisville, Kentucky. He will likely face twenty years after his trial."

Tom led the applause and thanked the lieutenant and Marshall Thompson. Then he invited them to join the breakfast. They agreed and thanked him.

After they had eaten, a siren sounded and a police van pulled up outside the tent. Cliffside's mayor stepped out, followed by three City Council members, a reporter, and photographer. Tom introduced them to those around the table and the audience that was gathering.

The mayor thanked them for the applause but said, "We are not the ones who deserve the applause. You, in my opinion, are the most generous civic-minded corporation I have ever known. Your Green Power Network corporation has organized and financed the people of Cliffside at a time of citywide emergency. We have experienced your corporate purposes in action. Area rescue boats manned by police, firemen, and EMT trucks in the eastern and western counties have been helping us for several days due to heavy rain. Your imaginative sandbags blocked the roaring brook so it wouldn't run down Hill Street on all your neighbors was a magnificent gesture. I have asked myself: These are not wealthy people, so how can they do this? Tom Carlton's family have been farmers in this community for a hundred-and-fifty years. Just farmers, not wealthy people.

The simple answer is a partnership between a farmer and his lifelong friend—the sun in the sky, who incidentally, is a friend to all of us. A partnership created a gift for us all: Agrivoltaic solar panels with growing vegetables underneath. All of this on a farm just up the hill, with forty-five acres of its farmland giving its power to all of us.

"Agrivoltaic is a protective influence for Cliffside, and a carbon free power source helping our climate survive."

Abbey and Phoebe were standing just on the edge of the mayor's group when his phone rang. Abbey excused himself and stepped a few paces away. It was his police colleague, back in the building behind a snack bar on the phone.

She said, "Our mic is working at Joe Hendrick's garage."

He asked, "Has Kirk Forest managed to get away from the police?"

The woman answered, "No, he's been arrested and is in our hands again. He'll be on his way to Louisville shortly. A Marshall named Robert Preston nabbed him. We have a helicopter coming in and landing at the City College ballfield to pick them up."

Abbey asked, "Now?"

She answered, "Yes."

Abbey was silent for a moment then said, "Call the chief of police, tell him what you just told me, and ask him to get to the Louisville DOJ immediately. Ask if there is a Marshall by the name of Robert Preston. I'll need an answer quickly, so ask him to call me."

At that moment, Abbey heard Robert Preston say, "He was waiting for a cab to pick him up. He's now handcuffed to me, and we're headed to the City College ballfield for a helicopter ride back to Louisville."

"I'll need them brought here first," Abbey said to the woman. She agreed.

Abbey whispered to Phoebe, "We have to hold the group together a short while longer. Slide over and check your holster—we may need it. Get Lieutenant Gibson to me quickly."

In the meantime, Robert Preston and Kirk Forest appeared along with some of the police.

Gibson moved quietly over to Abbey and asked, "Do you have your cuffs with you?"

Gibson nodded. Abbey said, "Good go stand next to Kirk Forest—when I nod, put your cuffs on his spare wrist while Phoebe draws her weapon."

Gibson said quietly, "He's already cuffed to Preston."

Abbey replied, "I know, but we may want them both. I'm waiting for the chief to call me back. I may want you to immediately cuff Kirk to yourself."

Gibson nodded and moved over close to Kirk.

Abbey's phone rang. The chief said, "Robert Preston was killed last year. You have a phony there chained to our arrested Kirk Forest."

Abbey nodded to Gibson and Phoebe. Gibson immediately locked his wrist to Forest. Preston tried hard to get to his weapon. Phoebe, with her gun drawn, was there first, and he couldn't quite make it.

The mayor shouted back to them, "What's going on?"

Abbey replied, "This Robert Preston is a phony. Like he told you, he has a helicopter waiting at the City College ballpark to whisk Kirk and himself back to Louisville. Mr. Mayor, we're just stopping their escape plans for our prisoner."

The chief of police called Abbey back and said, "An armed State Police helicopter is on its way and will deal with the helicopter at the City College ballfield."

The mayor took Tom Carlton by the hand and said, "You have created a very effective organization, and I'm proud to have you in our city with your Green Power Network. I hope you'll be successful and will reduce the fossil fuel sources of our power and replace it with green power."

The End

ACKNOWLEDGEMENTS

John Gergely: Champlain Office of Economic Opportunity (Coaching and Support)

Brian Dooley: Green Mountain Power Company

Don Fernandez: Research

Scott Connolly: Technical Consultant & Research

Valco of Vermont: A planner's report

The New York Times, by Ellen Rosen, published June 28th, 2022: "Can Dual-Use Of Solar Panels Provide Power and Share Space With Crops?"

The National Grid, by Jesse Stevenses

Grid-scale Energy Storage Development by Technology Type, Parts One and Two

OPB, by Monica Samayao, published June 7, 2022: "Oregon Utility Powers Up Nation's First Large-Scale Wind, Solar, and Battery Facility."

The Nation, by David McDermott Hughes, published May 23, 2022: "What if the Wind and Sunshine Really Belonged to All of Us?"

PV Magazine, by Ann Fischer, published June 1st, 2022: "Huge Battery Facility in Texas Goes Online."

BBC, by Matt McGrath, published July 4th, 2022: "Climate Change: 'Sand battery' Could Solve Green Energy's Big Problem."

Living on the Grid, by William L. Thompson

Fundamentals of Renewable Energy Processes, 2nd Edition, by Aldo Vieira da Rosa

The Grid, by Gretchen Bakke, PHD

Polar Night Energy: "Sand Battery"

USA Today, by Elizabeth Weiss, published February 6, 2023: "What Is Carbon Dioxide?"

USA Today, by Dennis Wagner, published October 13, 2022: "In Search of Lithium, Changing Our Future."

The Street, by Maxx Chatsko, published November 1, 2022: "National Gas Power Plants Begin Their Inevitable Decline."

Inside Climate News, by Dan Gearino, published January 20, 2023: "New Wind and Solar are Cheaper…"

USGS, published April 13, 2023: "The Potential for Hydrogen For Next-Generation Energy."

University of Arizona: Research into "What is Agrivoltaic"

Blue Wave Solar: Paul Knowlton's farm in Grafton, MA

Seven Days, by Kevin McCallum, published April 12, 2023: "Vermont Needs More Green Power…"

The University of Arizona, by Stacy Pigott, published September 2, 2019, "Agrivoltaics Proves Mutually Beneficial…"

Thank You